Exploiting structure in
non-convex quadratic optimization
and
gas network planning under
uncertainty

vorgelegt von
Dipl.-Math. Jonas Schweiger
geboren in Kassel

von der Fakultät II – Mathematik und Naturwissenschaften
der Technischen Universität Berlin
zur Erlangung des akademischen Grades

Doktor der Naturwissenschaften
– Dr. rer. nat. –

genehmigte Dissertation

Promotionsausschuss:

Vorsitzender: Prof. Dr. Jochen Blath
Gutachter: Prof. Dr. Thorsten Koch
Prof. Dr. Andrea Lodi

Tag der wissenschaftlichen Aussprache: 3. Februar 2017

Berlin 2017

Bibliografische Information der Deutschen Nationalbibliothek

Die Deutsche Nationalbibliothek verzeichnet diese Publikation in der
Deutschen Nationalbibliografie; detaillierte bibliografische Daten sind
im Internet über http://dnb.d-nb.de abrufbar.

ISBN 978-3-8325-4667-0

Logos Verlag Berlin GmbH
Comeniushof, Gubener Str. 47,
10243 Berlin
Tel.: +49 (0)30 42 85 10 90
Fax: +49 (0)30 42 85 10 92
INTERNET: http://www.logos-verlag.de

Abstract

The amazing success of computational mathematical optimization over the last decades has been driven more by insights into mathematical structures than by the advance of computing technology. In this vein, we address applications, where nonconvexity in the model and uncertainty in the data pose principal difficulties.

The first part of the thesis deals with non-convex quadratic programs. Branch-and-bound methods for this problem class depend on tight relaxations. We contribute in several ways: First, we establish a new way to handle missing linearization variables in the well-known Reformulation-Linearization-Technique (RLT). This is implemented into the commercial software CPLEX. Second, we study the optimization of a quadratic objective over the standard simplex or a knapsack constraint. These basic structures appear as part of many complex models. Exploiting connections to the maximum clique problem and RLT, we derive new valid inequalities. Using exact and heuristic separation methods, we demonstrate the impact of the new inequalities on the relaxation and the global optimization of these problems. Third, we strengthen the state-of-the-art relaxation for the pooling problem, a well-known non-convex quadratic problem, which is, for example, relevant in the petrochemical industry. We propose a novel relaxation that captures the essential non-convex structure of the problem but is small enough for an in-depth study. We provide a complete inner description in terms of the extreme points as well as an outer description in terms of inequalities defining its convex hull (which is not a polyhedron). We show that the resulting valid convex inequalities significantly strengthen the standard relaxation of the pooling problem.

The second part of this thesis focusses on a common challenge in real world applications, namely, the uncertainty entailed in the input data. We study the extension of a gas transport network, e.g., from our project partner Open Grid Europe GmbH. For a single scenario this maps to a challenging non-convex MINLP. As the future transport patterns are highly uncertain, we propose a robust model to best prepare the network operator for an array of scenarios. We develop a custom decomposition approach that makes use of the hierarchical structure of network extensions and the loose coupling between the scenarios. The algorithm used the single-scenario problem as black-box subproblem allowing the generalization of our approach to problems with the same structure. The scenario-expanded version of this problem is out of reach for today's general-purpose MINLP solvers. Yet our approach provides primal and dual bounds for instances with up to 256 scenarios and solves many of them to optimality.

Extensive computational studies show the impact of our work.

Zusammenfassung

Der bemerkenswerte Erfolg der angewandten mathematischen Optimierung in den letzten Dekaden ist mehr auf Einsichten in mathematische Strukturen zurückzuführen, als auf eine Steigerung der Rechenleistung. In diesem Sinne adressieren wir Anwendungen, in denen Nichtkonvexität und Unsicherheit in den Daten die Hauptschwierigkeiten darstellen.

Der erste Teil dieser Arbeit beschäftigt sich mit nichtkonvexen quadratischen Optimierungsproblemen. Relaxierungen sind integraler Bestandteil von Branch-and-bound-Lösungsmethoden für diese Problemkategorie. Wir leisten folgende Beiträge: Erstens beschreiben wir eine neue Art fehlende Linearisierungsvariablen, in der so genannten Reformulation-Linearization-Technique (RLT), zu behandeln. Diese wird inzwischen in der kommerziellen Software CPLEX verwendet. Zweitens beschäftigen wir uns mit der Optimierung einer quadratischen Zielfunktion über die Standardsimplex oder einen so genannten Knapsack-Constraint. Solche grundlegenden Strukturen sind Teil vieler komplexer Modelle. Wir benutzen bekannte Verbindungen zum maximalen Cliquenproblem sowie zu RLT, um neue gültige Ungleichungen herzuleiten, die die Relaxierung verstärken. Drittens beschäftigen wir uns mit dem Pooling Problem, das z.B. in der Erdölindustrie relevant ist. Wie leiten eine neue Relaxierung her, die die wesentliche nichtkonvexe Struktur des Problems erfasst, aber klein genug für eine grundlegende Untersuchung ist. Wir geben eine innere Beschreibung in Form der Extrempunkte, sowie eine äußere Beschreibung in Form von Ungleichungen, die die konvexe Hülle (welche im Allgemeinen kein Polyeder ist) beschreiben, an. Wir zeigen, dass neuen die Ungleichungen die Relaxierung des Pooling Problems erheblich verstärken.

Der zweite Teil der Arbeit befasst sich mit einer weiteren Herausforderung in realen Anwendungen, nämlich Unsicherheit in den Eingabedaten. Konkret untersuchen wir die Optimierung des Ausbaus eines Gastransportnetzes, wie z.B. von unserem Projektpartner Open Grid Europe GmbH. Dieses Problem ist bereits bei gegebenen Eingabedaten ein schweres nichtkonvexes gemischt-ganzzahliges Optimierungsproblem. Da zukünftige Nutzungsmuster des Netzes mit großer Unsicherheit behaftet sind, beschreiben wir ein robustes Modell, um den Netzbetreiber gegen verschiedene Szenarien abzusichern. Wir entwickeln einen speziellen Dekompositionsalgorithmus unter Berücksichtigung der hierarchischen Struktur der Ausbauten und der schwachen Kopplung zwischen den Szenarien. Unser Ansatz liefert primale und duale Schranken für Instanzen mit bis zu 256 Szenarien und löst viele beweisbar optimal.

Umfangreiche Rechnungen bestätigen die Effizient der vorgestellten Methoden.

Financial acknowledgments

This work was supported by

- the Research Campus Modal funded by the German Federal Ministry of Education and Research (fund number 05M14ZAM),

- the ICT COST Action TD1207 "Mathematical Optimization in the Decision Support Systems for Efficient and Robust Energy Networks",

- the EU Initial Training Network MINO (Grant Agreement Number 316647), and

- the DFG Collaborative Research Centre CRC/Transregio 154 "Mathematical Modelling, Simulation and Optimization Using the Example of Gas Networks" within projects B06 and Z01.

The support is strongly acknowledged.

Acknowledgments

Probably every PhD student knows this question: "So, which year of your PhD are you in?" Over time, the reaction to my answer shifted from "Oh, you still have time!" over "Oh, then you must be almost finished!" to "Oh, really? That's long!". In the end, it took more than six years and of course it was time to get it done, but I don't regret this long time. I had the chance to live in different countries, work for academic institutions with a different view on our field, get the feeling for a big company, and acquire a very broad and valuable set of skills. Experiences like this are not made alone; they are connected to people I met on the way. People that helped me, instructed me, challenged me, taught me lessons in all areas of life, professional as well as personal. People that are simply amazing to work and hang out with. This is the place to say "Thank you".

After finishing my undergrad studies, I started the "Oh, you still have time!" phase in the optimization department at Zuse Institute Berlin. I want to thank Martin Grötschel, Thorsten Koch, Ralf Borndörfer, and the entire group for creating and maintaining an inspiring and productive atmosphere. Every time I come back to the optimization floor it feels like coming home! I particularly thank my long-time roommate Robert Schwarz for all the interesting discussions, many liters of tea and for introducing me to the game of Go. I thank Timo Berthold for encouraging and advising me at all points in my career. Thank you, Thorsten, for the support over the years. Big thanks to the SCIP team for helping me debugging many times.

I thank my colleagues from the working group "Energy", Benjamin Hiller, Jesco Humpola, Thorsten Koch, Thomas Lehmann, Ralf Lenz, and Robert Schwarz for vital discussions and a nice working atmosphere. I thank the colleagues from the academic partners in the FORNE project, in particular Rüdiger Schultz for supporting and encouraging me during my work on network planning under uncertainty. I also thank Rüdiger for inviting me to Duisburg and Claudia Stangl and Uwe Gotzes for taking good care of me and showing me around. Thanks also to Klaus Spreckelsen from Open Grid Europe for nice dinner conversations about software development and gas network planning.

I am indebted to Andreas Löbel for the excellent support of the IT-infrastructure in the optimization department. I thank the ZIB library, especially Regina Kossick, for the great service they provide. The internet is definitely not able to replace a good librarian!

One day in the middle of the "Oh, then you must be almost finished!" phase Tobias Achterberg, who at this time was still working for IBM, told me about a position as "Experienced Researcher" in the CPLEX team within the MINO project. Thank you, Tobias, for opening this door to me!

So, after 4 years at ZIB I moved to Bologna and started doing research in the CPLEX team. I want to thank the entire CPLEX team for the warm welcome, the tight integration in the team and the support. Very special thanks to my direct co-workers Pierre Bonami, Andrea Lodi, and Andrea Tramontani. You taught me so much and it was an amazing experience to work with you! Thanks, Pierre, for answering my questions about MIP, QP, CPLEX code and tooling so patiently. I thank Andrea Lodi for supporting and advising me and especially for being part of my PhD committee. Special thanks go to Andrea Tramontani and my managers Giovanni Palumbo and Xavier Nodet for all the support they gave me. I can only imagine how many exceptions from IBM rules were made for me as you always protected me from the pitfalls of IBM bureaucracy and cleared my way.

I thank the OR group at the university of Bologna and its guests at the time, specially Andrea Baggio, Andrea Bettinelli, Claudio Gambella, Alberto Santini, Dimitri Thomopulos, Paolo Tubertini, and Sven Wiese. Thanks for the pleasant atmosphere, many dinners, and in particular for agreeing not to go to Garisenda anymore! Special thanks to Sven for the nice time we had in the office at CIRAM, during coffee breaks and beers after work.

The MINO project also sent me to Paris. I thank Claudia D'Ambrosio for making this stay very productive, for the nice time and joint work at Anticafe, and for integrating me in the project about the pooling problem, where I had the chance to work also with Jeff Linderoth and Jim Luedtke. Thanks, Jim, for the nice remote working sessions over skype. You helped me out several times! Thanks, Jeff, for an unforgettable time in Sevilla. Thanks CJJ for all the fun in this collaboration.

After 18 month my time at IBM was over, I was definitely in the "Oh, really? That is long!" phase, but my PhD was still unfinished. Fortunately, Frauke Liers had pointed me to the graduate school of the Collaborative Research Center TRR154 which gave me a scholarship. Thank you, Frauke, for having me on the list and supporting me on the last meters of my PhD. Without you, I probably would have never finished the paper on multi-scenario gas network planning. I thank Alexander Martin and his entire group for the nice and productive atmosphere, especially during the stay in Kleinwalsertal.

Very special thanks go to Anne Ain, Kati Ain, Timo Berthold, Andreas Eisenblätter, Lena Hupp, Craig Kershaw, Benjamin Müller, and Felipe Serrano for reading parts of this thesis and providing valuable comments.

I deeply thank my family and especially my parents, Claudia and Hartmut, and my stepparents, Ute and Axel, for all the love and support. Unfortunately Oma Helene, Opa Siegfried, Opa Helmut, and Opa Gustl will not be able to celebrate with me, but I am sure they would have been very proud. Danke, Oma Irma, für die Liebe, die du mir in all den Jahren gegeben hast.

Last, but not least, I thank all my friends who made sure I never forget there is also a life outside of mathematics and research.

Danke!

Contents

1 Introduction

Probably every mathematician experienced the typical reactions after saying that we do math for a living. Two of them are "Oh, I always hated math!" and "Oh, that is sooo interesting! Math is everywhere!". While the first terminates the discussion about mathematics quite quickly, the second one is often source of a nice conversation. There is a third reaction that is remarkable: "Oh, you must be a genius with numbers. Can you compute [Any meaningless quotidian quantity] in your head?" This reaction brings two misconceptions about mathematics to the surface. The first is that mathematics is about numbers while it is actually about abstractions and structures. The second is that *doing mathematics* actually means *computing* something. Profound knowledge of the mathematical structures, however, is the key to design algorithms that perform practical computations in the most efficient way. The amazing success of mathematical optimization in the last decades has been driven by a thorough exploitation of such structures, which is why solver software for so-called linear and mixed-integer linear programs routinely computes optimal solutions for practical problems with hundreds of thousands of variables [Bix+00].

In this thesis, we consider nonlinear and non-convex problems and exploit their structure for more efficient computations. Convexity in the objective function and in the feasible set is a strong property with wide implications, but many applications demand non-convex models. Solution methods for convex and non-convex problems are very different and in this thesis we exclusively focus on algorithms for non-convex problems, so-called *Mixed-Integer Nonlinear Programs* *(MINLP)*. Branch-and-bound is the algorithmic scheme employed by most state-of-the-art software packages [ANT; BAR; Cou; SCI] to solve non-convex MINLP. In this scheme, the non-convex problem is relaxed to a problem that can be efficiently solved. The problem is then iteratively split into subproblems to be able to construct tighter and tighter relaxations. The algorithm ensures a globally optimal solution or proves that no feasible solution exists. In Chapter 2 we provide formal definitions of MINLP and its subclasses and describe the branch-and-bound algorithm.

Tight and efficiently solvable relaxations are crucial for good performance and an active field of research. In general-purpose MINLP solvers, typically non-convex functions are replaced by appropriate linear or convex over- and underestimators. While for each elementary function these approximations are often best possible, the combination of functions and the interaction between different aspects of the model are not taken into account and lead to weak relaxations. Using the problem structure to tighten the relaxations of specific classes of non-convex problems is one topic of this thesis.

In the first part of the thesis, we concentrate on quadratic programs as an important subclass of MINLP. In this problem class, the multiplication of two variables is the only source of nonlinearity and McCormick [McC76] already in 1976 provided a tight relaxation for the product of two bounded variables. This termwise relaxation, however, is weak when terms interact or when more constraints are present in the model. The so-called *Reformulation-Linearization-Technique (RLT)* [SA92] is well known to strengthen the relaxation of quadratic programs by capturing structure of the constraints. In a nutshell, the ideas is to multiply a linear constraint by a variable which yields product terms. Auxiliary variables representing the products of two variables are used to reformulate the constraint making it linear again. Different implementations for this method have different ways to handle product terms that appear only in constraints generated by this approach. We propose to project them out by replacing them with appropriate over- and underestimators, an approach that has not been described in the literature to the best of our knowledge. Within the work on this thesis, the author implemented the projected RLT cuts in the commercial solver CPLEX [IBMb], where they are enabled in the default settings for optimization problems with a non-convex quadratic objective in version 12.7.0. An overview about reformulations and relaxations for quadratic programs together with a presentation of (projected) RLT is given in Chapter 3.

In Chapter 4, we consider the first specific quadratic problem with the objective to strengthen its relaxation by using its particular structure. The task is the optimization of a (generally non-convex) quadratic objective function over the standard simplex. The standard simplex is a polytope in $\mathbb{R}^d$ which is described by nonnegative variables $x \geq 0$ and the linear constraint

$$\sum_{i=1}^{d} x_i = 1.$$

A problem of this type is called *Standard Quadratic Program (SQP)* [Bom98]. SQP has fundamental relations to other problems, e.g., Copositive Programming [Dür10]. The well-known theorem of Motzkin-Straus [MS65] states that the problem of finding the size of a maximum clique in a graph can be formulated as a Standard Quadratic Program. This connection to the combinatorial maximum clique problem allows us to derive a valid inequality for SQP for each graph on d nodes. For bipartite graphs we show connections to RLT and derive a stronger and more general class of valid inequalities. Exact and heuristic separation algorithms are proposed and implemented. An extensive computational study shows that the inequalities considerably strengthen the McCormick relaxation and yield a significant improvement in terms of dual bound and time to optimality. The proposed inequalities can be adapted to sets described by an inequality constraint with more general coefficients. Such inequalities and the standard simplex are important substructures in many optimization models which makes our inequalities widely applicable.

In Chapter 5, the *Pooling Problem* is the second application we consider in this thesis. The pooling problem is a classical non-convex optimization problem with a wide range of applications for example in the petrochemical [QG95] or the mining industry [Bol+15]. In short, the task is to route flow from inputs over pools to outputs. The catch is that the material has attributes whose concentrations are known at the inputs, but are constrained at the outputs. The challenge is to route the flow through the network such that the quality constraints at the outputs are met, taking into account that the material is blended at the pools and at the outputs. The blending of the material differentiates the pooling problem from other multi-commodity flow problems and give rise to non-convex quadratic constraints. After reviewing the classical formulations and relaxations of this problem, we use the state-of-the-art pq-formulation as starting point to derive a novel relaxation for the problem. This is done by reducing the network to its minimal essence comprising only an aggregated version of the inputs, one output, one pool and one material attribute. At the same time, the relaxation still captures parts of the central non-convex structure of the model. Finding an appropriate relaxation which is tight enough to improve the pq-formulation, small enough for a thorough study, and rich enough in the essential structure of the problem is the first contribution in this chapter. After finding the relaxation, we follow the route that has been walked by integer linear programmers for decades, namely to study the extreme points of the relaxation and translate the inner description of the set into an outer description based on the defining inequalities. The set of interest is in general not a polyhedron and has infinitely many extreme points. However, a finite subset of the extreme points leads to a parametrized set of linear inequalities that are used to derive new valid linear and nonlinear inequalities for the pooling problem. The inequalities are added directly or by means of the separation of gradient inequalities to the pq-formulation. A computational study shows that the proposed relaxation is tight enough to strengthen the McCormick relaxation of the pq-formulation. Especially on sparse instances, the additional inequalities provide a significant speed-up of the global solution of the problem and allow to solve several instances that are not solved by the pq-formulation. This completes our work on strong relaxation for quadratic programs.

The second part of this thesis treats the challenge of handling uncertainty in gas network extension planning. We start this part in Chapter 6 by briefly sketching the model and the algorithmic approach for deterministic gas network planning. Gas networks consist of passive elements such as pipes and active elements such as control valves and compressors. The behavior of the network is described in terms of continuous variables for physical quantities such as pressure and flow and discrete variables for operational decisions of active devices. The nonlinear relationship between pressure and flow in the network makes the problem a challenging non-convex MINLP even for one scenario.

In practice, the input data of the problem is seldom known exactly. The reasons for uncertainty in the input data are manifold and include measurement errors, model simplifications, and uncertain forecasts. Investments into the gas network infrastructure are extremely costly and impact the network performance for decades to come. Therefore, small changes in the quality of the decisions can result in substantial financial gains or losses. At the same time the future transport patterns in Europe are highly uncertain as the EU is increasingly dependent on gas imports [Com]. While deterministic planning approaches focus on one bottleneck scenario, the long planning horizon paired with high uncertainty ask for planning methods that take several scenarios for future demand into account in order to prepare the network for future challenges. We use the framework of *Robust Optimization* [BGN09] where instead of assuming that the data that describes the objective and the constraints is known, the input data assumed to realize itself within an uncertainty set. The decisions that are to be determined then must be *robust*, i.e., they need to be feasible no matter how the data manifests itself with the uncertainty set. The concept of robust optimization is reviewed in Section 2.3. For linear mixed-integer problems, tractable robust counterparts can be derived for several classes of uncertainty sets, such as conic or polyhedral sets. As we are facing a complex MINLP, much less is known about tractable robust counterparts. We therefore consider a discrete uncertainty set that consists of a finite number of scenarios for the uncertain data. This reflects the situation in which different scenarios are collected from historical data, future forecasts, or domain experts. One typical property of these models is that only very few variables and constraints couple the different scenarios and thus decomposition methods are a common weapon of choice. We also follow this route and design a decomposition algorithm that takes the particular structure of our model into account.

We address this problem in Chapter 7. We model the gas network extension problem for several demand scenarios as two-stage robust problem where the investment decisions are in the first and the operational decisions in the second stage. We seek network extensions from a given candidate set that render all scenarios feasible at the lowest possible investment cost. The two-stage nature of the problem stems from the fact that the investment decisions have to be taken jointly for all scenarios, while the decisions how to operate the scenarios are taken on a per-scenario basis. One feature of the model is that the extension candidates naturally form a hierarchy. For each extension candidate several variants are available where the "bigger" extension includes all features of "smaller" ones. Consider for example some point in the network where an active element might be constructed. With the first level in the hierarchy being the option to build nothing, the second is to build a valve that can be opened and closed. The next level, the compressor, adds an "active" state that allows to increase the pressure in the direction of the flow, but also has "open" and "closed" states and thus has strictly more features than the valve, but at higher cost.

To tackle this problem, we propose a scenario decomposition and a tailored branch and bound algorithm that exploits the particular structure of the problem. The network extension problem for the individual scenarios is used as a sub-problem whose solutions guide the search. We propose several ways to avoid re-solving the same scenario when this would not generate new information. Furthermore, we develop effective heuristics to find feasible solutions. We perform a computational study on realistic network topologies from the publicly available gaslib [Gasa]. The arising MINLP in a scenario-expanded formulation of the robust problem have up to 360 000 variables and 330 000 constraints from which 70 000 are nonlinear equations. This large problem size and the complexity of the model are the reasons why this problem in our experience is out of scope for today's general-purpose MINLP solvers. Our approach is able to provide primal and dual bounds and solves a large range of instances to proven optimality.

The work on gas transport network planning in this second part of the thesis was done in the ForNe project[1]. The aim of ForNe was to address several challenges Europe's larges gas network operator Open Grid Europe GmbH (OGE) faces in their daily operation, as well as tactical and strategical planning and bring the findings into practice by developing software solutions. ForNe was a joint project between Open Grid Europe and several academic institutions, among them the Zuse Institute Berlin, and ended in 2015. The project was awarded the EURO Excellence in Practice Award 2016[2] for its achievements documented in

---

[1] http://www.zib.de/projects/forne-research-cooperation-network-optimization
[2] https://www.euro-online.org/web/pages/1595/eepa-winners-2016

the book [Koc+15].

The use of automated mathematical optimization has high potential to result in considerable savings. Powerful methods for deterministic planning gas networks have been developed in the FORNE project. They are used as subroutine in this work. According to Open Grid Europe, simultaneously optimizing for several demand scenarios is the next step that has the potential to change the way network planning is done in the industry [Spr16]. The development of such an approach was therefore one of the advanced goals of the FORNE project. An implementation of theses methods has been used by OGE.

Publications and collaborations   Significant parts of this thesis have emerged from collaborations with colleagues from all over the world and have been published in peer-reviewed conference proceedings or have been submitted to international journals for publication.

In detail, this includes:

- The work on projected RLT inequalities in Section 3.3 and Standard Quadratic Programming presented in Chapter 4 was done while the author was working within the EU Initial Training Network MINO[3] for IBM CPLEX Optimization in Bologna and is a collaboration with Pierre Bonami from IBM Spain, Andrea Lodi who was at the University of Bologna at this time and is now at École Polytechnique de Montréal, and Andrea Tramontani from IBM Italy. A publication that contains large parts of Section 3.3 and Chapter 4 has been submitted to SIAM Journal on Optimizationand is available as preprint [Bon+16].

- The author started the work on the pooling problem in a 2-month secondment at the École Polytechnique in Paris which was also part of the MINO project. The work then continued while the author was at the graduate school of the collaborative research center TRR 154. It is a collaboration with Claudia D'Ambrosio from École Polytechnique Paris and Jim Luedtke and Jeff Linderoth from the University of Wisconsin-Madison. A publication that will contain large parts of Chapter 5 is in preparation.

- The models and methods on deterministic network extension planning in Chapter 6 have been developed in the FORNE project and credits go to numerous colleagues from all partners in the FORNE project. Chapter 6 is given for completeness and not considered a genuine contribution of this thesis.

---

[3]http://cordis.europa.eu/project/rcn/105341_en.html

- The model and the solution approach for topology optimization of gas networks under uncertainty in Chapter 7 have been developed within the ForNe project while the author was at Zuse Institute Berlin. The implementation was completed with support of the graduate school of the collaborative research center TRR 154. A short paper that was published in the Operations Research Proceedings 2014 [Sch16] and a journal version that was coauthored by Frauke Liers from the University of Erlangen-Nürnberg and is submitted to the journal "Optimization and Engineering" contains large parts of Chapter 7. Both publications contain also parts of Chapter 6 for completeness.

2 Concepts

This section briefly introduces the basic concepts that are used throughout this thesis. In Section 2.1 the different classes of mathematical programs are defined. We focus on optimization problems whose objective function and feasible set is described by closed-form algebraic expressions. Some or all of the variables are allowed to have the requirement that they can only take integer values. Solution methods for such *Mixed-Integer Nonlinear Programs* are sketched in Section 2.2.

As last concept, *Robust Optimization* as a framework to handle uncertainty in the input data of an optimization problem is briefly introduced in Section 2.3.

2.1 Classes of mathematical programs

In mathematical optimization the task is to find the extreme value (the biggest or the smallest value possible) with respect to some objective function given constraints on the possible solution values, i.e., problems of the form

$$\max/\min g(x) \text{ subject to } x \in \mathcal{X}.$$

We note that each maximization problem can be reformulated into a minimization problem by replacing the objective $\max g(x)$ by $-\min -g(x)$. For ease of exposition we assume minimization in the following.

More precisely, we consider *Mixed-Integer Nonlinear Programs* (MINLP), i.e., optimization problems of the following form:

$$
\begin{align}
\min \quad & g_0(x) & & \text{(2.1a)} \\
\text{s.t.} \quad & g_i(x) \leq 0 & \text{for all } i \in \mathcal{M} & \text{(2.1b)} \\
& \underline{x_j} \leq x_j \leq \overline{x_j} & \text{for all } j \in \mathcal{N} & \text{(2.1c)} \\
& x_j \in \mathbb{Z} & \text{for all } j \in \mathcal{I} & \text{(2.1d)}
\end{align}
$$

where $\mathcal{N} := \{1, \ldots, n\}$ is the index set of the variables, $\mathcal{I} \subseteq \mathcal{N}$ the index set of the integer variables, $g_i : \mathbb{R}^n \to \mathbb{R}$ for $i \in \{0\} \cup \mathcal{M}$, $\mathcal{M} := \{1, \ldots, m\}$, and $\underline{x_i} \in \mathbb{R} \cup \{-\infty\}$ and $\overline{x_i} \in \mathbb{R} \cup \{+\infty\}$ are lower and upper bounds on the variables, respectively. The function $g_0(x)$ is called *objective function*, the functions $g_i(x)$

constraint functions and the inequalities $g_i(x) \leq 0$ are referred to as *constraints*. We assume all functions are factorable, i.e., can be reformulated as sums of products of univariate functions possibly involving auxiliary variables. The vectors $\underline{x}$ and $\overline{x}$ are called *bound vectors* and the hyperrectangle $[\underline{x}, \overline{x}] = [\underline{x_1}, \overline{x_1}] \times [\underline{x_2}, \overline{x_2}] \times \cdots \times [\underline{x_n}, \overline{x_n}]$ is the *bounding box*. For a given MINLP the set

$$\mathcal{X} := \left\{ x \in \mathbb{R}^n \mid g_i(x) \leq 0 \text{ for all } i \in \mathcal{M}, x \in [\underline{x}, \overline{x}], x_j \in \mathbb{Z} \text{ for all } j \in \mathcal{I} \right\}$$

is called *set of feasible solutions*. Let $c^* \in \mathbb{R} \cup \{\pm\infty\}$ be defined by

$$c^* := \inf \left\{ g_0(x) \mid x \in \mathcal{X} \right\}.$$

If c^* is finite, then we call it *optimal solution value*, otherwise we call the MINLP *infeasible* if $c^* = +\infty$ or *unbounded* if $c^* = -\infty$. All minimizers of the objective function are called *(global) optimal solutions*. Last, we call an MINLP a *feasibility problem* if the objective function g_0 is constant.

MINLP is a very general framework and a lot of problems can be formulated in this way. As seen above maximization problems can be reformulated by minimizing the negative objective function, $\geq$ inequalities by multiplying the inequality by -1 to obtain an equivalent $\leq$ inequality and equations by a pair of opposite inequalities. We assume that all functions are factorable which ensures that they can efficiently be stored and evaluated by an *expression tree*. See [Vig13] for a mathematically precise definition of factorability and its consequences.

In many definitions, the objective function is assumed to be linear. A nonlinear objective $g_0(x)$ function can then be modeled by adding one additional variable z which is to be minimized and the constraint $g_0(x) \leq z$.

Subclasses Depending on the structure of the functions and the number of integer variables, several subclasses of MINLP are distinguished. First, we focus on the continuous problems which have no integer variables ($\mathcal{I} = \emptyset$) and affine linear and quadratic functions. An affine linear function f can be written was $f(x) = a^T x - b$ where $a \in \mathbb{R}^n$ is a column vector and $b \in \mathbb{R}$ a scalar. A constraint is called linear it its constraint function is affine linear. A system of linear inequalities is typically written in matrix notation as $Ax \leq b$ where $A \in \mathbb{R}^{m \times n}$ is called *constraint matrix* and $b \in \mathbb{R}^m$ is called *right hand side vector*. Quadratic functions are more general by allowing the multiplication of two variables and have the form $f(x) = x^T Q x + ax + b$ where $Q \in \mathbb{R}^{n \times n}$ is the coefficient matrix, $a \in \mathbb{R}^n$ is the vector of linear coefficients and $b \in \mathbb{R}$ is a scalar. A constraint of the form

$$\|Ax + b\|_2 \leq c^T x + d$$

is called *second order cone (SOC) constraint*. The points that fulfill

$$2x_1x_2 \geq x_3^2 + \cdots + x_k^2, \ x_1, x_2 \geq 0$$

describe a k-dimensional *rotated second order cone*. Second order cone constraints are convex quadratic constraints and indeed every convex quadratic constraint can be written as SOC constraint. Sets that allow a representation as intersection of second order cones (possibly involving auxiliary variables) are called *second order cone representable*.

With these definitions the following continuous problems have specific names:

- A problem with affine linear objective function and linear constraints is called *Linear Program* (LP).

- A problem with quadratic objective function and linear constraints is called *Quadratic Program* (QP).

- A problem with quadratic objective function and quadratic constraints is called *Quadratically Constrained Quadratic Program* (QCQP).

- A problem with linear objective and linear and second order cone constraints constraints is called *Second Order Cone Program* (SOCP).

- All continuous programs not falling in any of the previous categories are called *Nonlinear Program* (NLP).

We name subclasses that are relevant for this thesis while there are of course many more in the literature. QPs and QCQPs play a special role in this thesis and Chapter 3 studies them in greater depth.

Often these programs appear together with integer requirements on some or all variables. The terms *mixed-integer* expresses the possibility that some or all variables have integer requirements. In the literature, mixed-integer is sometimes also meant in a strict sense, where some but not all variables have integer requirements, i.e., $\mathcal{N} \neq \mathcal{I} \neq \emptyset$. Programs where all variables have integer requirements, i.e., $\mathcal{I} = \mathcal{N}$, are called *(pure) integer* and programs with $x_i \in \{0,1\}$ for all $i \in \mathcal{N}$ are called *(pure) binary*. Often these attributes are combined with the continuous problem category and we speak about Mixed-Integer Nonlinear Programming (MINLP) and alike. However, not all combinations are common. For historical reasons Integer Linear Programs (ILP) and Mixed-Integer Linear Programs (MILP) are often referred to a IP and MIP, respectively.

Relaxations Relaxations are a powerful tool to get *dual* bounds of MINLP. Formally, a mathematical program R is a relaxation of another mathematical program P if every point that is feasible in P can be translated into a feasible point in R and the value of the optimal solution of R is smaller or equal to the optimal solution of P. Often, but not always, P and R have the same objective function; take the Lagrangean Relaxation (e.g., [Rus06]) where some constraints are dualized into the objective, as an example. The feasible set of R is typically larger than the one of P, but is structurally more well-behaved and allows to be solved more efficiently than P.

One source of great complication are integer variables. Relaxing the integrality condition yields the *continuous relaxation* which often can be computed efficiently. Relaxations are at the core of most exact solution approaches as they provide bounds on the global solution and are therefore essential to prove that the optimal solution has been found.

Convexity Convexity is a powerful property of sets and functions. A set X is called convex if for $x_1, x_2 \in X$ and for all $\lambda \in [0,1]$, the point $\lambda x_1 + (1 - \lambda)x_2 \in X$. A function $f : X \to \mathbb{R}$ is called *convex* on a convex domain $X \subset \mathbb{R}^n$ if for all $x_1, x_2 \in X$ and for all $\lambda \in [0,1]$ the following inequality holds:

$$f(\lambda x_1 + (1 - \lambda)x_2) \leq \lambda f(x_1) + (1 - \lambda)f(x_2) \tag{2.2}$$

A function f is called *concave* if $-f$ is convex.

Numerous equivalent definitions of convexity exist. One additional one is in terms of the function's epigraph. The graph $\mathrm{gr}(f)$ of f is defined by

$$\mathrm{gr}(f) := \{(x,y) \in \mathbb{R}^n \times \mathbb{R} \mid f(x) = y\}$$

and the epigraph $\mathrm{epi}(f)$ of f consists of all points on or above the graph

$$\mathrm{epi}(f) := \{(x,y) \in \mathbb{R}^n \times \mathbb{R} \mid f(x) \leq y\}$$

A function is convex if and only if its epigraph is a convex set.

As linearity, convexity allows to derive global information about a function from local information. An affine linear function is completely characterized by its gradient and some shift vector. In contrast to a linear function whose gradient is constant on its domain, the gradient of a convex function takes several values. Second order derivatives provide a criterion for convexity of a twice continuously differentiable function as such a function is convex if the Hessian is positive-semidefinite, i.e., has no negative Eigenvalues. For a one-dimensional function, linearity means that the function is completely described by a constant first-order derivative (slope) and a shift. One-dimensional convex functions are those with the second derivative, if it exists, greater or equal to 0 such that the slope can only increase.

In MINLP, convexity is a desirable property because if both the feasible region and the objective function are convex every local optimum is already globally optimal. A point $x^\star$ is called *local optimum* if there exists some $\epsilon > 0$ such that $g_0(z) \geq g_0(x^\star)$ for every feasible z with $\|z - x^\star\| < \epsilon$. Clearly, every global optimum is also a local optimum, but if the feasible region and the objective function are convex the reverse is also true and local optima are also globally optimal. To highlight the importance of convexity, an MINLP whose continuous relaxation has a convex objective and convex feasible set is called *convex* MINLP while general MINLP are called *non-convex*.

In practice, convex NLP can be solved efficiently by interior-point methods. For LP, QP, and QCQP these algorithms have proven polynomial runtime. For details on convexity analysis and convex optimization we refer to [Roc70; BV04]. The global solution of non-convex NLP is also called *global optimization* and offers a rich literature, e.g., the books [LM06; HPT00; LS13].

2.2 Introduction to MINLP

Branch-and-bound [LD60] is the dominating algorithmic framework to solve MILPs and MINLPs. Following the divide-and-conquer principle, the problem is repeatedly split into smaller problems with additional constraints which make the problems easier to solve. Let us first review the branch-and-bound algorithm for mixed-integer linear problems before extending it to MINLP.

2.2.1 LP-based branch-and-bound for MILP

Consider a MILP. The idea is to solve the continuous relaxation to get a lower bound. Let $\tilde{x}$ be a relaxation solution. If $\tilde{x}$ is integral, i.e., all integer variables take integer values, then the solution is feasible and optimal for the problem. Otherwise the problem is split into two disjoint problems in such a way, that every feasible solution is in one of the two subproblems, but the current solution to the relaxation is not. Typically, a disjunction from an integer variable is used to split the problem. Let $i \in \mathcal{I}$ be such that $\tilde{x}_i \notin \mathbb{N}$. Then two subproblems each with one of the constraints $x_i \leq \lfloor \tilde{x}_i \rfloor$ and $x_i \geq \lceil \tilde{x}_i \rceil$ are created. Clearly, $\tilde{x}$ is not included in any of them, but all feasible solutions are contained in either of them. This splitting into several subproblems is called *branching*. Solvers have sophisticated *branching rules* to select the variable to branch on. The same procedure is then repeated recursively to the subproblems.

The subproblems form a tree, where the original problem is the root node and every branching adds two children. Subproblems that have not been solved yet are collected in a queue of so-called *open nodes* and *node selection rules* are responsible for selecting the next node to be solved. The relaxation value of a node determines the node's *dual bound* as the best solution value that can be obtained in the subtree below the node. The node in the queue with minimum relaxation value determines the *global dual bound* for the problem which is increasing as nodes get treated because the dual bounds of the children are always greater or equal to the dual bounds of the parent. The series of dual bounds are complemented by series of *primal bounds*, i.e., objective values of feasible solutions. Apart from the relaxation, solvers use a variety of heuristics to find feasible solutions and improve the primal bound. The current best known feasible solution is called *incumbent*.

The *bounding* step is the key trick for this to form an efficient algorithm since it prevents the exploration of the entire tree. In this step, the dual bound of a given node is compared to the current primal bound. If the dual bound of the node is greater or equal than the primal bound, it is clear that no improving solution can be contained in the subtree that is encoded in the node and thus the node (and with it the entire subtree) can be *pruned*.

While this describes the general framework, a state-of-the-art solver has many tricks to improve the solution process; cutting planes to tighten the relaxation, bound propagation and probing to tighten variable bounds, a large set of divers heuristics to find good feasible solution, to name just the most important ones.

2.2.2 Convexification

One straigt-forward generalization to MINLP would be to replace LP-relaxations by the respective NLP-relaxations. While this was indeed proposed for convex MINLP [GR85], it is impractical for general MINLP. For general MINLP the NLP-relaxation is still non-convex and thus in general as hard to solve as the MINLP itself. To simplify the presentation, we assume in the following that MINLPs are formulated with a linear objective function.

To avoid the solution of non-convex relaxation, the problem is again relaxed to a convex NLP which can be solved efficiently. This relaxation gives a valid lower bound to the non-convex NLP-relaxation and also the the original MINLP. The relaxation of a non-convex set by a convex one is referred to as *convexification*.

The best convexification would of course be the convex hull $\mathrm{conv}(\mathcal{X})$ of the feasible set $\mathcal{X}$. however, it is difficult to compute in most practical applications. A more viable way to derive a convex relaxation is to convexify each constraint individually. For a constraint $g_i(x) \leq 0$, what is needed is a *convex underestimator* for $g_i(x)$. A convex underestimator of a function g_i on a subset S of the function's domain is a convex function f_i with

$$f_i(x) \leq g_i(x) \qquad\qquad \text{for all } x \in S. \qquad (2.3)$$

Factorability of g_i ensures that convex underestimators can always be computed if convex understimators for the univariate functions are known [McC76; Vig13].

While in principle every convex underestimator can be used, the one with the highest values gives the best approximation. This gives rise to the definition of the *convex envelope* on a set S as the best convex estimator, i.e., a convex underestimator $Vex_S(g_i)$ with

$$Vex_S(g_i)(x) \geq f_i(x) \qquad\qquad \text{for all } x \in S \qquad (2.4)$$

for all convex underestimators $f_i(x)$ of $g_i(x)$. The convex envelope can also be defined in terms of the epigraph of $g_i(x)$ since it is the function whose epigraph equals the convex hull of the epigraph of $g_i(x)$ on S.

Note, the dependence of convex underestimators and envelopes on the set S. On a subset of S, less points have to fulfill the conditions (2.3) and (2.4) such that tighter estimators can be constructed and therefore the convex envelope on the subset might be higher. As an example take the sine function. On its entire domain $\mathbb{R}$, the convex envelope is

$$Vex_{\mathbb{R}}(\sin)(x) = -1.$$

On the interval $[\pi, 2\pi]$, however, the function is convex and therefore its own convex envelope

$$Vex_{[\pi,2\pi]}(\sin)(x) = \sin(x).$$

In general, the computation of convex envelopes is a very challenging task and it has been shown in [Tar04] that even evaluating the convex envelope at a single point might be as difficult as solving the original MINLP. We refer to [Roc70] for a more formal treatment of convex analysis in general and convex envelopes in particular.

When the problem contains also $\geq$-inequalities, the terms *concave overestimator* and *concave envelope* $Cav_S(g_i)$ are common and defined analogously.

Sometimes the computation of the convex envelope can be carried out for the different summands that define a function independently. If the function $g_i(x)$ is representable as sum of functions over some index set $\mathcal{K}$

$$g_i(x) = \sum_{k \in \mathcal{K}} g_{ik}(x)$$

and the convex envelope of $g_i(x)$ is the sum of the convex envelopes of the summands

$$Vex_S(g_i) = \sum_{k \in \mathcal{K}} Vex_S(g_{ik})(x)$$

then $g_i(x)$ is called *sum decomposable*.

Convexification of different functions have been studied in the literature, e.g., [MF05; BST09; McC76; TS02; TS04; Tar04; Tar08; AMF95; CL12; BMN11] to name only a few. With αBB [AMF95; Adj+98; AAF98] a general approach has been proposed where the idea is to relax every non-convex function by adding a negative convex function in such a way that the sum is convex.

The convexification of the individual functions disregards the interaction between the different constraints. With the study of the feasible set, it is often possible to considerably improve the convex relaxation. This will be the general topic of Chapters 4 and 5 where the structure of the constraint sets for Standard Quadratic Programs and Pooling problems are studied.

2.2.3 Branch-and-bound for MINLP: Spatial Branching

Using the convexified NLP-relaxation as relaxation, we can extend the LP-based branch-and-bound algorithm for MILP to an algorithm for general MINLP. The general framework is the same and only briefly sketched here. For a more detailed description, we refer to the recent surveys [Bel+13; Vig13; VG16].

Like for MILP, a relaxation is solved to get lower bounds, heuristics are employed to find feasible solutions, branching on integer variables with fraction relaxation values is done to reduce integer infeasibility. In addition, the convex relaxations are updated according to bound changes in the variables which are involved. This is often done by separation, where (like with general-purpose cutting planes for MILP) parts of the relaxation are cut off by adding additional constraints. The main difference arises at nodes where the relaxation is integer feasible. If it is also feasible for the non-convex constraints, then the solution is feasible and indeed optimal for the node. Otherwise, branching comes as a last resort. A branching variable $x_i \in \mathcal{N}$ is chosen that participates in a non-convex function (more precisely in a non-convex expression) and two nodes are created, one with the additional constraint $x_i \leq \bar{x}_i$ and the other with $x_i \geq \bar{x}_i$. However, in contrast to branching on integer variables taking fractional values, the relaxation solution is not excluded by branching directly, but by the ability to tighten the convex relaxation due to the tighter bounds. This form of branching which will generally happen on continuous variables is called *spatial branching*. See Fig. 6.2 on page 146 for an example of spatial branching. Figure 6.2a shows the linear relaxation of the equation $y = |x|x$ on its original interval. Next, Fig. 6.2b shows the relaxations in the two child nodes after branching on $x = 0$ in the same plot.

Branch-and-bound with spatial branching is implemented in all state-of-the-art solvers for non-convex MINLP. BARON [BAR; Sah96; TS05], Couenne [Cou; Bel+09] and SCIP [SCI; VG16; Vig13] use linear underestimators and thus solve LP-relaxations. ANTIGONE [ANT; MF14; MF13; MSF15] uses piecewise linear underestimators and solves a MILP-relaxation. CPLEX [IBMb; BBL14] solves a convex QP-relaxation to solve non-convex MIQPs. See [BV11] for a nice review on available software packages for MINLP.

Of course, a sophisticated solver implements all kinds of tricks to make the algorithm work effectively. As for MILP, heuristics play a major role in finding good feasible solutions, e.g., [Ber14; BG14; DAm+10; AH14; Aud+04]. In the context of MINLP techniques to reduce the volume of the bounding box are of particular importance since tight bounds allow tighter underestimators, e.g., [Bel+09; Gle+16].

2.3 Robust Optimization

In the area of *Robust Optimization* models and methods are established to handle uncertainty in the input data of optimization problems. This section is meant to be a short introduction into the principle ideas of robust optimization. For a more complete and mathematical description as well as practical guides on the use of robust optimization, we refer to [BGN09; BBC11; GYH15] and the references therein.

The above review of MINLP assumes that all the problem data is known at the time the model is built and solved. Often enough, this is not the case. This might be because the data cannot be measured exactly or because it refers to data which only can be known in the future. Traditionally, the problem of uncertain data is tackled by replacing it by appropriate expected values or forecasts, but it is evident that this can lead to poor or even infeasible decisions for the true realization of the data.

One paradigm to treat data uncertainty is *Robust Optimization* [BGN09]. The principle idea is that the constraint functions that describe an instance of an optimization problem are parametrized by another parameter. The parameter represents the uncertainty in the data, so its true value is unknown. However, it is known to belong to a so-called *uncertainty set*. The desired solution then has to be *robust* against this uncertainty set, which means that it needs to be feasible no matter how the uncertain data manifests itself within the uncertainty set. Robust optimization therefore protects against the worst-case realization of the uncertain data. The task is to find a robust solution with optimum guaranteed cost.

Mathematically, we can think of the functions in (2.1) to depend on an uncertain parameter $\omega \in \Omega$, where Ω is the uncertainty set. With the same reformulation argument that we used for MINLP, we can restrict ourselves to uncertain data in the constraints and a known objective function. The model is then

$$\min \quad g_0(x) \tag{2.5a}$$

$$\text{s.t.} \quad g_i(x, \omega) \leq 0 \qquad \qquad \text{for all } i \in \mathcal{M}, \ \omega \in \Omega. \tag{2.5b}$$

Typically, the structural properties (e.g., linearity, convexity, etc.) of the functions $g_i(x, \omega)$ remains the same for different realizations of ω. In principle, the robust problem is then a semi-infinite one with finitely many variables but an infinite number of constraints as the uncertainty set is not necessarily a discrete set.

The big practical relevance of robust optimization emerges as in many cases the semi-infinite optimization problem can be reformulated into a compact and "tractable" equivalent. Compact and tractable in the situation means that the reformulation is of polynomial size and that the resulting robust counterpart is in the same complexity class as the original problem. As an example take the problem studied in the seminal paper [BN99], namely linear programming with an ellipsoidal uncertainty set. Without going into the technical details here, the robust counterpart is an SOCP, i.e., a convex quadratic problem, that can be solved in polynomial time.

Modeling a problem in a robust way typically results in very conservative decisions as feasibility is also required for very unlikely corner cases in the uncertainty set and no prioritization between very likely and very unlikely realizations takes place. Note, that the uncertainty set is part of the model input. Sensible modeling of the uncertainty set is therefore key for the practical success of robust optimization as small uncertainty sets lead to insufficient protection and large uncertainty sets make the problem very conservative and often result in very expensive solutions. Initially mostly polytopal and ellipsoidal uncertainty sets were used to ensure a certain variability in the input parameters of the problem. Since then, a lot of research into lowering the "price of robustness", i.e., the penalty one has to pay in order to make a model robust, has been done and tractable Robust Counterparts are known for a lot of types of uncertainty sets. Most prominently, we want to mention Γ-uncertainty by Bertsimas and Sim [BS03;

BS04], who also coined the descriptive term "price of robustness".

In Chapter 7, a robust extension of a non-convex MINLP is proposed. For linear mixed-integer problems, tractable robust counterparts can be derived for several classes of uncertainty sets, such as conic or polyhedral sets. As we are facing a complex MINLP, much less is known about tractable robust counterparts. Therefore, we see robust optimization as a framework to model uncertainty in the input data and restrict ourselves to uncertainty in the form of a discrete set of scenarios and develop a custom solution approach based on decomposition.

2.3.1 Multistage Robust Optimization

In classical robust optimization, the decisions are to be taken *here-and-now*, that is, decisions have to be taken before the uncertain data reveals itself. The decisions have to be taken simultaneously for all realizations of the uncertain parameter and no adjustment is allowed. In practice, often some of the variables correspond to *wait-and-see* decisions that only have to be taken after the outcome of uncertainty is known. Moreover, uncertainty sometimes reveals gradually as hence the decisions can take a certain amount of revealed information into account. In this case, the decisions for a multistage process, where the here-and-now decisions are in the first and the wait-and-see decision are in the later stages. For example, in the gas network application from Chapter 7 the task is to decide on network extensions under the restriction that a discrete set of demand scenarios can be operated in the extended network. There, the decisions which extensions should be built are of here-and-now type and correspond to the first stage. Given the network extensions, the operational decisions have to be taken for each scenario independently. The operational decisions are thus of wait-and-see type and belong to the second stage. We restrict this introduction to two stages.

Mathematically, the variables can then be grouped into first stage variables x and second stage variables $y(\omega)$ which depend on the scenario. The model then can be written as

$$\min \quad g_0(x) \tag{2.6a}$$
$$\text{s.t.} \quad g_i(x, y(\omega), \omega) \leq 0 \qquad \text{for all } i \in \mathcal{M}, \ \omega \in \Omega \tag{2.6b}$$
$$f_i(x) \leq 0 \qquad \text{for all } i \in \mathcal{M}'. \tag{2.6c}$$

(2.6c) models constraints that act only on the first stage variables. The constraints (2.6b) constrain the second stage variables and ensure that for every realization ω and the first stage decisions x, a feasible second stage decision $y(\omega)$ exists.

For continuous uncertainty sets, the Robust Counterpart (2.6) has infinitely many variables and constraints and even if a finite reformulation exists, the problem is often intractable [Ben+04]. If the uncertainty set is a finite set, then 2.6 is of course a finite problem. However, due to the typically large number of scenarios and thus very large problem size, it is not very amendable to solution by general-purpose solvers. Notice that for fixed first stage decisions x, the problem decomposes as $y(\omega)$ can be computed for each scenario individually and no constraints connect different scenarios. Decomposition methods, such as Benders decomposition, therefore often come to the rescue. As the resulting subproblems for the gas network application are non-convex MINLP, we present in Chapter 7 a decomposition approach which is tailored for the specific situation.

After this short introduction, we want to mention that accommodating robust multi-stage optimization problems is an active research field and several different approaches have been proposed, among them *Adjustable Robust Optimization* [Ben+04] and *Recoverable Robust Optimization* [Lie+09].

3 Reformulations and relaxations for quadratic programs

The study and solution of mathematical programs that involve only linear and quadratic functions is known as *Quadratic Programming*. While the step from linear to quadratic might appear small at first, quadratic functions are much more general in terms of modeling. As an example take the condition that a variable x is binary, i.e., can take only the values 0 or 1. This condition is equivalently formulated by the quadratic constraint $x = x^2$. In

Quadratic programs can be divided into two classes: Convex and non-convex. Convex quadratic problems can be solved in polynomial time [KTK80; BV04]. The picture changes for non-convex problems. Theoretically, the problem of optimizing a (in general non-convex) quadratic objective function over a finite set of linear inequalities is $\mathcal{NP}$-hard [PV92; Sah74; Vav90]. Notwithstanding, many instances of non-convex quadratic problems can be solved in practice [BBL14; Ber+12].

This chapter describes the algorithmic framework used in the two proceeding chapters. As the problems considered in this thesis are non-convex, we focus on solution methods for non-convex problems. After introducing the necessary notation in Section 3.1, we review standard methodology to obtain convex relaxation for non-convex quadratic problems in Section 3.2. In Section 3.3 we introduce the Reformulation-Linearization-Technique (RLT), which is a standard technique to strengthen the relaxation of a quadratic program, and proposed a way to handle quadratic terms that only occur in RLT constraints by projection (Section 3.3.1). Finally, we comment on the impact of the implementation of the projected RLT inequalities by the author in the commercial solver CPLEX.

3.1 Definitions and notation

A function f is called quadratic if it has the form $f(x) = x^T Q x + a x + b$ where $Q \in \mathbb{R}^{n \times n}$ is a symmetric matrix. Quadratic programming thus deals with this special NLP:

$$\min \quad x^T Q_0 x + a_0 x + b_0 \tag{3.1a}$$
$$\text{s.t.} \quad x^T Q_i x + a_i x + b_i \le 0 \qquad \text{for all } i \in \mathcal{M} \tag{3.1b}$$
$$\underline{x_j} \le x_j \le \overline{x_j} \qquad \text{for all } j \in \mathcal{N} \tag{3.1c}$$

Finite bounds are needed for those variables that appear in products of two variables as otherwise a convex relaxation cannot be computed. For simplicity, we assume finite bounds on all variables, i.e., $\underline{x}, \overline{x} \in \mathbb{R}^n$. Furthermore, we restrict the exposition to continuous problems without integer variables to focus on the treatment of the quadratic objective function and constraints.

If both, the objective function and at least one constraint function are quadratic functions ($Q_i \ne 0$ for $i = 0$ and at least one $i \in \mathcal{M}$) then Problem 3.1 is called *Quadratically Constraint Quadratic Program* (QCQP). The optimization of a quadratic objective function over a polyhedron is known as *Quadratic Program* (QP) while programs where the feasible region is defined by quadratic inequalities and a linear objective function is known as *Quadratically Constrained Program* (QCP). Of course QPs and QCQP can be described as QCPs. The class of QCPs is very broad and problems such as integer programs with bounded variables and polynomial optimization can be reformulated as QCPs.

As mentioned in the previous chapter, convexity is a key property for the design of solution methods for MINLP and quadratic programs in particular. One definition of convexity says that a twice continuously differentiable function is convex if and only if its Hessian is positive semi-definite. As the Hessian of a quadratic function $f(()x) = x^T Q x + a x + b$ is the matrix Q, f is convex if and only if Q is positive semi-definite.

To simplify the presentation of this chapter, we will considers QCPs and treat QPs as special case wherever this is appropriate. The feasible set of the QCP is denoted by

$$\mathcal{X} := \left\{ x \in \mathbb{R}^n \,\middle|\, x^T Q_i x + a_i x + b_i \le 0 \text{ for all } i \in \mathcal{M}, x \in [\underline{x}, \overline{x}] \right\}$$

Quadratic programs and quadratic expressions can be associated with an undirected graph $G = (V, A)$ which has the variables as nodes and an arc $a = \{i, j\}$ if the product x_i, x_j appears with nonzero coefficient in the program or expression. G is called *support graph*. The elements of the matrix are often thought of as edge-weights, i.e., Q_{ij} being the weight of edge $\{i, j\}$. A term $x_i x_j$ is called *quadratic* if $i = j$ and *bilinear* if $i \ne j$. An expression that contains only bilinear terms is called *bilinear expression*.

3.2 Convexification

When facing a non-convex problem, convex relaxations are of major importance. Of course $\text{conv}(\mathcal{X})$ would be the best convexification, but it is typically impractical to compute. Instead convex under- and overestimators are used and the ability to compute good over- and underestimator is key for solution methods like spatial branch-and-bound. In this section we will discuss the arguably most commonly used technique to construct convex relaxations for bipartite expressions: The termwise McCormick relaxation. Therefore each product of two variables is replaced by an additional variable that is bounded by the under- and overestimators. For quadratic terms gradient hyperplanes and secants are used. For bilinear terms McCormick [McC76] described appropriate under- and overestimators. The McCormick relaxation has been studied in great depth and in the next subsection we will review the most relevant results for the problems treated in this thesis. Among alternative convexification schemes are αBB (e.g., [Adj+98; Adj+96; AAF98; AMF95] and the SDP-relaxation (e.g., [BV08] and the references therein) which we do not cover here in detail. See [Ans12] for a comparison of different convexification schemes.

3.2.1 McCormick relaxation

Termwise linearization of all bilinear and quadratic terms is the most common way to convexify QCPs. To this end an auxiliary variable Y_{ij} is introduced for each term $x_i x_j$ in the model together with the constraint

$$Y_{ij} = x_i x_j. \tag{3.2}$$

Quadratic terms $Q_{ii} x_i^2$ are over- and underestimated using gradient and secant hyperplanes that might be separated during the solution process. For bilinear terms, the equation $Y_{ij} = x_i x_j$ is non-convex and can be convexified by the so-called *McCormick* inequalities.

The mathematical statements involve some matrix notation which we quickly introduce here. The *outer product* xx^T of x describes the matrix of quadratic and bilinear terms

$$xx^T = \begin{pmatrix} x_1 x_1 & x_1 x_2 & \cdots & x_1 x_n \\ x_2 x_1 & \ddots & & \vdots \\ \vdots & & & \\ x_n x_1 & \cdots & & x_n x_n \end{pmatrix}. \tag{3.3}$$

In this compact notation, constraint (3.2) then reads

$$Y := xx^T. \tag{3.4}$$

For two matrices $Q, Y \in \mathbb{R}^{n \times n}$ we define the *matrix scalar product* $\langle Q, Y \rangle$ as

$$\langle Q, Y \rangle := \sum_{i=1}^{n} \sum_{j=1}^{n} Q_{ij} Y_{ij}.$$

Problem 3.1 is then reformulated by replacing every occurrence of the term $x_i x_j$ by Y_{ij} or equivalently every occurrence of $x^T Q_i x$ by $\langle Q_i, Y \rangle$. The reformulated problem is then

$$\begin{align}
\min \quad & \langle Q_o, Y \rangle + a_0 x + b_0 & & \text{(3.5a)} \\
\text{s.t.} \quad & \langle Q_i, Y \rangle + a_i x + b_i \leq 0 & \text{for all } i \in \mathcal{M} & \text{(3.5b)} \\
& \underline{x_j} \leq x_j \leq \overline{x_j} & \text{for all } j \in \mathcal{N} & \text{(3.5c)} \\
& Y = xx^T. & & \text{(3.5d)}
\end{align}$$

Notice that the objective function (3.5a) and the constraints (3.5b) are now linear and all the nonlinearity is moved into the constraint (3.5d).

In practice, the matrix equality (3.5d) is only enforced for those indices (i, j) where the corresponding product $x_i x_j$ appears somewhere in the model as on all other entries, violations don't influence the optimal solution or its value. Furthermore, linearization variables are only really added to the model for indices $i \leq j$ and symmetric product identified.

The most common way to form a convex relaxation of Problem 3.5 is to relax each non-convex equality $Y_{ij} = x_i x_j$ separately. We refer to this approach as *(termwise) McCormick relaxation*. Quadratic terms $Y_{ii} = x_i^2$ can be relaxed by using gradient inequalities underestimating x_i^2 and the secant inequality as overestimator. The infinitely many gradient inequalities can be separated in cutting plane loop. For bilinear terms $Y_{ij} = x_i x_j$, $i \neq j$, the so-called McCormick inequalities provide under- and overestimators [McC76]:

$$\begin{align}
\overline{x_j} x_i + \overline{x_i} x_j - \overline{x_i}\, \overline{x_j} &\leq Y_{ij}, & \text{(3.6)} \\
\underline{x_j} x_i + \underline{x_i} x_j - \underline{x_i}\, \underline{x_j} &\leq Y_{ij}, & \text{(3.7)} \\
\overline{x_j} x_i + \underline{x_i} x_j - \underline{x_i}\, \overline{x_j} &\geq Y_{ij}, & \text{(3.8)} \\
\underline{x_j} x_i + \overline{x_i} x_j - \overline{x_i}\, \underline{x_j} &\geq Y_{ij}. & \text{(3.9)}
\end{align}$$

Clearly, the McCormick inequalities depend on the bounds on the variables and get tighter as the bounds get tighter.

Another way to view (3.6)–(3.9) is in terms of under- and overestimators. The expression

$$\underline{m}_{x_i x_j}(x_i, x_j) := \max(\overline{x_j} x_i + \overline{x_i} x_j - \overline{x_i}\, \overline{x_j},\ \underline{x_j} x_i + \underline{x_i} x_j - \underline{x_i}\, \underline{x_j}) \tag{3.10}$$

is a convex underestimator of a product $x_i x_j$ and

$$\overline{m}_{x_i x_j}(x_i, x_j) := \min(\overline{x}_j x_i + \underline{x}_i x_j - \underline{x}_i \overline{x}_j, \; x_j x_i + \overline{x}_i x_j - \overline{x}_i \underline{x}_j) \tag{3.11}$$

is a concave overestimator of $x_i x_j$. A bilinear expression $g(x) = x^T Q x$ with $Q_{ii} = 0$ is then underestimated by

$$\underline{m}_g(x) := \sum_{\substack{(i,j) \in \mathcal{N} \times \mathcal{N} \\ Q_{ij} > 0}} Q_{ij} \underline{m}_{x_i x_j}(x_i, x_j) + \sum_{\substack{(i,j) \in \mathcal{N} \times \mathcal{N} \\ Q_{ij} < 0}} Q_{ij} \overline{m}_{x_i x_j}(x_i, x_j) \tag{3.12}$$

and overestimated by

$$\overline{m}_g(x) := \sum_{\substack{(i,j) \in \mathcal{N} \times \mathcal{N} \\ Q_{ij} > 0}} Q_{ij} \overline{m}_{x_i x_j}(x_i, x_j) + \sum_{\substack{(i,j) \in \mathcal{N} \times \mathcal{N} \\ Q_{ij} < 0}} Q_{ij} \underline{m}_{x_i x_j}(x_i, x_j) \tag{3.13}$$

We refer to $\underline{m}_g(x)$ and $\overline{m}_g(x)$ as *McCormick estimators* of g. Replacing all non-convex quadratic terms in Problem 3.1 by McCormick underestimators is equivalent the termwise McCormick reformulation, but has the drawback of being only piecewise linear.

Exactness of McCormick After McCormick introduced the relaxation in [McC76], it was proved in [AF83] that (3.6)–(3.9) indeed describe the convex hull of the set

$$\{(x_i, x_j, Y_{ij}) \in [\underline{x}_i, \overline{x}_i] \times [\underline{x}_j, \overline{x}_j] \times \mathbb{R} \mid Y_{ij} = x_i x_j\}.$$

This does not mean that the McCormick relaxation provides the convex hull of the feasible set of a given QP since each term is relaxed individually but the interaction between the variables is not captured. While $Y = x x^T$ has to hold for each feasible solution, the McCormick inequalities generally only give a coarse relaxation of Problem 3.5 or more precisely of its feasible region $\mathcal{X}$. The quality of the relaxation is difficult to assess in general due to the various interactions of the functions that define $\mathcal{X}$.

However, it is known under which conditions the McCormick envelopes are the convex and concave envelopes and the termwise McCormick is the convex hull of the graph of a bilinear function. Let $g(x) = x^T Q x$ be a bilinear function, i.e., $Q_{ii} = 0$ for $i \in \mathcal{N}$, and define its graph by

$$\mathcal{X}(g) = \{(x,y) \in \mathbb{R}^n \times \mathbb{R} \mid \underline{x} \leq x \leq \overline{x}, \; y = g(x)\}$$

and the termwise McCormick relaxation by

$$\mathcal{M}(g) = \{(x,y) \in \mathbb{R}^n \times \mathbb{R} \mid \underline{x} \leq x \leq \overline{x}, \; y = \langle Q, Y \rangle, \; x, Y \text{ fulfill (3.6)–(3.9)}\}.$$

Of course it would be desirable that $\text{conv}(\mathcal{X}(g)) = \mathcal{M}(g)$ as it is always the case of $n = 2$, but generally $\mathcal{M}(g)$ is a relaxation of $\text{conv}(\mathcal{X}(g))$ and it only holds that

$$\text{conv}(\mathcal{X}(g)) \subset \mathcal{M}(g). \tag{3.14}$$

In terms of under- and overestimators, this means that the McCormick estimators $\underline{m}_g(x)$ and $\overline{m}_g(x)$ do not necessarily provide the convex and concave envelopes.

Some sufficient conditions for equality in (3.14) have been know for a while, but the question of when the McCormick underestimator this is the case was answered in a recent paper [MSF15]. We rephrase the result to match our notation:

Theorem 3.1 ([MSF15]): *The McCormick underestimator $\underline{m}_g(x)$ is the convex envelope of bilinear expression $g(x) = x^T Q x$ if and only if there does not exist a cycle C_n of length n in the support graph with an odd number of positively-weighted edges Q_{ij}.*

Since bipartite graphs have no odd cycles, the following corollary follows directly:

Corollary 3.2 (e.g., [Cop+99; LNL12]): *If the support graph of an expression is bipartite and the coefficients are nonnegative, then the McCormick underestimator is the convex envelope.*

Another immediate consequence is that the McCormick relaxation provides the convex hull of an expression if there are no cycles.

Corollary 3.3: *If the support graph of an expression g has no cycles, then*

$$\text{conv}(\mathcal{X}(g)) = \mathcal{M}(g).$$

Proof. By Theorem 3.1, the McCormick underestimator is the convex envelope of the expression. As the negative of every convex underestimator of $-g$ is a concave overestimator for g and $-g$ also has no cycles, the McCormick overestimator is the concave envelope. As the McCormick under- and overestimators describe the convex and concave envelopes of g, respectively, the McCormick relaxation describes the convex hull of g. □

Error bounds on the termwise McCormick relaxation for bilinear expressions have been proved in [LNL12; Bol+16].

Several attempts have been made to improve the McCormick relaxation. The convex and concave envelopes of a bilinear expression g on a rectangular domain is known to be *vertex polyhedral* (e.g., [Rik97]). A function is called vertex polyhedral over a polyhedral domain if its epigraph is a polyhedron whose vertices correspond to the vertices of the domain. In this case each point on the graph of the function can be expressed as convex combination of vertex points $(x^i, g(x^i))$

of the graph of g, where $x^1, x^2, \ldots, x^{2^n}$ are the vertices of the bounding box $[\underline{x}, \overline{x}]$. In mathematical terms this reads

$$\text{gr}(g) = \left\{ (x, y) \in [\underline{x}, \overline{x}] \times \mathbb{R} \,\middle|\, x = \sum_{i=1}^{2^n} \lambda_i x^i, \; y = \sum_{i=1}^{2^n} \lambda_i g(x^i), \; \lambda \in \mathbb{R}^n_{\geq 0}, \; \sum_{i=1}^{2^n} \lambda_i = 1 \right\}$$

By means of LP-duality, this representation has been used to separate cutting planes from $\text{conv}(\mathcal{X}(g))$ [BST09]. [MF05] provides a construction algorithm for the convex envelope of vertex polyhedral functions in three variables which is used in the GloMIQO solver [MSF15]. [BMN11] describe the convex hull of a bilinear term, with the additional requirement that a bound on the value of the product is given, i.e., the convex hull of the set

$$\left\{ (x_i, x_j, Y_{ij}) \,\middle|\, Y_{ij} = x_i x_j, \; Y_{ij} \leq \alpha, \; x_i \in [\underline{x_i}, \overline{x_i}], \; x_j \in [\underline{x_j}, \overline{x_j}] \right\}.$$

In this case, the convex hull is not polyhedral.

3.2.2 Convex relaxations by matrix decomposition

The idea is to decompose the matrix Q in two matrices where one is positive semidefinite and the other is indefinite or negative semidefinite. In other words, the matrix is split into a convex and a non-convex part. The objective is to keep convex substructures in the relaxation as to make it as strong as possible. The non-convex remainder still needs to be convexified.

In mathematical terms, we consider the quadratic function

$$g(x) := x^T Q x. \tag{3.15}$$

We omit possible linear and constant terms for notational convenience. Assume the matrix $Q \subset \mathbb{R}^{n \times n}$ decomposes into $Q = S + P$ where $S, P \in \mathbb{R}^{n \times n}$ and $S \succeq 0$ is positive semidefinite. The function can then be written as

$$\begin{aligned} g(x) &= x^T Q x \\ &= x^T (S + P) x \\ &= x^T S x + x^T P x. \end{aligned}$$

The first term is convex by construction and is treated in the relaxation directly. The second might only be convex if the constraint was convex to start with. In the following we assume that the original problem was non-convex and thus that P is indefinite or negative semidefinite. The term $x^T P x$ then needs convexification, e.g., by the termwise McCormick relaxation.

The decomposition $Q = S + P$ is by no means unique. Several desirable properties of such a decomposition should be traded off: First, the relaxation should be as strong as possible. On the other hand, if $x^T P x$ is convexified by the McCormick relaxation, the matrix P should be as sparse as possible as to introduce only few linearization variables. In practice also the effort to compute the decomposition should be taken into account.

In the following we describe the Q-space reformulation as it is implemented in the commercial solver CPLEX in a little more detail. This is the reformulation we assume in Chapter 4. The Q-space reformulation is a very straightforward approach. It looks at all the terms $Q_{ij}x_i x_j$ and puts convex terms in S and non-convex ones in P. Since a term is convex if and only if it is constant ($Q_{ij} = 0$) or $i = j$ and $Q_{ij} \geq 0$, the matrices S and P and then defined by

$$S_{ij} = \begin{cases} Q_{ij} & \text{if } i = j \text{ and } Q_{ij} \geq 0 \\ 0 & \text{otherwise} \end{cases}$$

$$P_{ij} = \begin{cases} 0 & \text{if } i = j \text{ and } Q_{ij} \geq 0 \\ Q_{ij} & \text{otherwise.} \end{cases}$$

S describes a convex quadratic function since it is the sum of convex functions.

The products that have a nonzero in the non-convex part P are relaxed by the termwise McCormick relaxation. As before, the matrix $Y = xx^T$ is introduced and the problem is reformulated (again with the understanding that the linearization variable Y_{ij} is only introduced of $P_{ij} \neq 0$ for the objective of some constraint). The termwise McCormick relaxation is then a convex problem. The reformulation is than tackled by spatial branch-and-bound.

This reformulation is very effective in practice and CPLEX in its default settings for non-convex QP mostly relies on it [BBL14; Bon]. Another advantage is that the relaxation can be strengthened by RLT (see Section 3.3). Recently with Boolean Quadric Polytope (BQP) cuts [BGL16] another class of inequalities has been proposed that strengthen this formulation and CPLEX applies them by default [IBMa].

CPLEX allows for a second reformulation called *Eigenvalue-reformulation* [SBL10; FLM13] which is favorable when then problem is almost convex [Bon]. This approach uses the Schur-decomposition

$$Q = \sum_{i=1}^{n} \lambda_i v_i v_i^T$$

where λ_i are the eigenvalues and v_i the respective eigenvectors. Q is then split into a positive semidefinite and a negative semidefinite matrix by taking only the terms corresponding to positive and negative eigenvalues, respectively. Also GloMIQO uses this reformulation under certain conditions [MF13]. They use the term *eigenvector projections* for it. Several alternative decompositions have been studied in [FLM13].

3.3 Reformulation-Linearization Technique (RLT)

The major drawback of the termwise McCormick relaxation is that each term is relaxed independently, disrespecting the interactions of various constraints. The Reformulation-Linearization Technique (RLT) [SA92] is a way to strengthen the relaxation that combines several aspects of the model to create new valid constraints. The four McCormick inequalities (3.6)–(3.9) can be derived by this technique.

RLT consists of two steps. In the first step, valid constraints are multiplied by other constraints or by variables yielding an equation or inequality with higher order terms. In the second step, these terms are reformulated using linearization variables to obtain a linear constraint. The result is a valid constraint on the linearization variables that is often a very strong cutting plane [MF13].

Consider for example valid bound constraints for variables x_i and x_j:

$$\overline{x_i} - x_i \geq 0 \tag{3.16}$$

$$\overline{x_j} - x_j \geq 0 \tag{3.17}$$

The product of the left hand side terms of (3.16) and (3.17) is then also greater or equal to zero:

$$(\overline{x_i} - x_i)(\overline{x_j} - x_j) \geq 0. \tag{3.18}$$

This can then rearranged into

$$-\overline{x_j}x_i - \overline{x_i}x_j + \overline{x_i}\,\overline{x_j} + x_ix_j \geq 0. \tag{3.19}$$

In the linearization step, the identity $Y_{ij} = x_ix_j$ is used to linearize (3.19)

$$-\overline{x_j}x_i - \overline{x_i}x_j + \overline{x_i}\,\overline{x_j} + Y_{ij} \geq 0. \tag{3.20}$$

Equation (3.20) is then the McCormick constraint (3.6).

Instead of the simple bound constraints, two arbitrary linear constraints can be multiplied like this. If both are equations, then the result can be strengthened to be an equation as well.

The last variation is the multiplication of a linear constraint by a variable. We review it here in more details because this case is most relevant for Chapters 4 and 5. Consider a linear constraint

$$\sum_{i \in \mathcal{N}} a_i x_i \leq b \tag{3.21}$$

and a nonnegative variable x_j. Multiplying (3.21) with x_j yields the valid inequality

$$\sum_{i \in \mathcal{N}} a_i x_i x_j \leq b x_j \tag{3.22}$$

and reformulation gives the valid RLT inequality

$$\sum_{i \in \mathcal{N}} a_i Y_{ij} \leq b x_j \tag{3.23}$$

The same procedure is applicable to $\geq$ inequalities and equations. If the sign of the variable is not known, i.e., $\underline{x_j} < 0$ and $\overline{x_j} > 0$, then the constraint has to be an equation or the equation has to be multiplies by one of the bound constraints. If the variable is nonpositive, then the sign of the inequalities (3.22) and (3.23) has to be inverted. The constraints (3.23) are very strong and render many RLT inequalities from multiplying two linear constraints redundant.

By repeatedly applying this procedure, a hierarchy of relaxations using higher order terms can be established. For linear binary programs, this hierarchy has be shown to converge towards the convex hull [SA94]. We restrict this exposition to the first order that involves only bilinear and quadratic terms.

3.3.1 Projected RLT inequalities

The RLT constraints are known to be strong, but they might use linearization variables for products $x_i x_j$ that don't appear in the model. Typically, linearization variables (together with gradient, secant and McCormick inequalities and the possible need to branch on them to ensure the equation $Y_{ij} = x_i x_j$) are only introduced for those products which have a nonzero coefficient in one of the matrices Q_0 or Q_k, $k \in \mathcal{M}$. This poses a problem to the applicability of RLT when the RLT constraint uses terms that are not linearized. One possibility is to simply not apply the respective RLT constraint. GloMIQO in recent versions also adds linearization variables just to formulate strong RLT constraints [MSF15]. We propose to relax the RLT constraint by projecting out the variables corresponding to missing terms.

Precisely, let the set V_j collect all indices i, for which a linearization variable Y_{ij} exists

$$V_j = \left\{ i \in V \mid (Q_k)_{ij} \neq 0 \text{ for some } k \in \mathcal{M} \cup \{0\} \right\}. \tag{3.24}$$

Terms $x_i x_j$ for which $i \notin V_j$ are replaced by linear over- and under-estimators, i.e., linear functions $o_{ij}(x_i, x_j)$ and $u_{ij}(x_i, x_j)$ such that

$$o_{ij}(x_i, x_j) \geq x_i x_j,$$
$$u_{ij}(x_i, x_j) \leq x_i x_j.$$

Then, in the example above the projected RLT inequality from (3.23) is

$$\sum_{i \in V_j} a_i Y_{ij} + \sum_{\substack{i \notin V_j \\ a_i > 0}} a_i u_{ij}(x_i, x_j) + \sum_{\substack{i \notin V_j \\ a_i < 0}} a_i o_{ij}(x_i, x_j) \leq b x_j. \tag{3.25}$$

The McCormick inequalities are natural under- and overestimators, but they add additional nonzeros to the inequality. In the implementation in Chapter 4 we under- and overestimate x_i by its bounds such that the sparsity pattern of the projected RLT inequalities remains the same. Most over- and under-estimators become stronger as the bounds become tighter and (3.25) can be separated in the nodes of the branch-and-bound tree as locally valid cuts with estimators taking into account the local variable bounds, especially after branching.

Within the work on this thesis, the author implemented the projected RLT cuts in the commercial solver CPLEX [IBMb], where it is enabled in the default settings for optimization problems with a non-convex quadratic objective in version 12.7.0. Compared to the previous version 12.6.3, projected RLT is responsible for 11 additionally solved instances and a reduction on the runtime by 10 % in the shifted geometric mean over all non-convex QP instances in the internal CPLEX testset that are solved within a timelimit of 10000 seconds by at least one of the two versions. This number increases to 29 % considering only models that are affected by the separation of theses inequalities. Projected RLT is particularly helpful on hard instances. On instances where at least one version takes at least 1000 seconds to solve the problem to optimality the shifted geometric mean of the runtime is reduced 84 %.

4 Motzkin-Straus inequalities for Standard Quadratic Programming and generalizations

Standard quadratic programs are non-convex quadratic programs with the only constraint that variables must belong to a simplex. Even though standard quadratic programs essentially only have one linear constraint, this structural information can be used to considerably improve the McCormick relaxation. The applicability of RLT to this constraint is widely known and significantly improves the relaxation as we will confirm in the computational results in this chapter.

However, we go one step further. By a famous result of Motzkin and Straus, those problems are connected to the clique number of a graph. In this chapter this connection is used to derive strong cutting planes for standard quadratic programs that drastically tighten the relaxation. We derive in particular cuts that correspond to an underlying complete bipartite graph structure. We study the relation between these cuts and the classical ones obtained by the first level of the reformulation-linearization technique. By studying this relation, we derive a new type of valid inequalities that generalize both types of cuts and are stronger.

Finally, we show how to generalize the cuts to non-convex quadratic knapsack problems, i.e., to attack problems in which the feasible region is not restricted to be a simplex. This makes the proposed cuts a very versatile tool to tighten the relaxation of quadratic programs.

This chapter is organized as follows: In Section 4.1 we review the connection between standard quadratic programs and the clique number of a graph. In Section 4.2 we review the Q-space relaxation and the application of RLT for this problem. The first class of new valid inequalities based on the theorem of Motzkin-Straus is introduced in Section 4.3 and connections to RLT inequalities are shown. We call these inequalities *Motzkin-Straus Clique inequalities* (MSC inequalities for short). For complete bipartite graphs we provide an alternative RLT-based proof. Extending this result, in Section 4.4 we propose a second type of new inequalities that generalize both the RLT methodology and some of our previous inequalities. Consequently, we call these inequalities *generalized MSC*

33

bipartite inequalities (GMSC bipartite inequalities for short). Section 4.5 describes exact and heuristic methods to separate and strengthen violated inequalities. In Section 4.6 we present extensive computational results using the different cutting planes we propose in the context of spatial branch-and-bound. We show that our cuts allow to obtain a significantly better bound than reformulation-linearization cuts and reduce computing times for global optimality. In Section 4.7 we show how the cuts can be generalized if x is not required to be in the Standard Simplex, but fulfills a more general inequality. Finally, Section 4.8 concludes the chapter.

This chapter is based on joint work with Pierre Bonami, Andrea Lodi and Andrea Tramontani. Most of this chapter is contained in a paperthat is currently under review by SIAM Journal on Optimization and available as preprint [Bon+16].

4.1 Introduction

In this chapter, we study the problem of optimizing a quadratic function over the *standard simplex*, namely

$$\min \left\{ x^T Q x \,\middle|\, x \in \Delta \right\}, \tag{StQP}$$

where the standard simplex is defined as

$$\Delta = \left\{ x \in \mathbb{R}^d \,\middle|\, \sum_{i=1}^{d} x_i = 1, x \geq 0 \right\},$$

and $Q \in \mathbb{R}^d \times \mathbb{R}^d$ is a symmetric matrix. We do not make any further assumption on Q and the optimization problem (StQP) is a non-convex optimization problem, being generally referred to as *Standard Quadratic Program* [Bom98]. Variants or generalizations of StQP appear in many applications where the sum of fractions has to sum up to 1 or where exactly one of several (binary) options has to be chosen. Applications in finance and the Quadratic Assignment problem are just two examples. Problem StQP also has fundamental relations with copositive programming [Dür10]. In particular, StQP has an exact reformulation as a copositive programming problem [Bom+00] and the solution of StQP can be used to test if a matrix is copositive [BEJ16].

Although StQP is a purely continuous optimization problem it has strong connections with combinatorial optimization and in particular with the maximum clique problem by a remarkable result of Motzkin and Straus [MS65]. Below, we remind the definition of the maximum clique problem and state formally this result.

A clique in a simple, undirected graph $G = (V, E)$ is a subset of nodes where every node is connected to all other nodes. The size of the largest clique in G is called *clique number of* G and denoted by $\omega(G)$. The problem of computing the clique number is one of Karp's 21 NP-hard problems [Kar72].

The Motzkin-Straus Theorem connects the clique number of a graph with StQP.

Theorem 4.1 (Motzkin-Straus [MS65]): *Let A be the adjacency matrix of a simple, undirected graph $G = (V, E)$ and $\omega(G)$ its clique number. Then, the following relation holds:*

$$\max \left\{ x^T A x \mid x \in \Delta \right\} = 1 - \frac{1}{\omega(G)}.$$

Note the identification of variables x_i with nodes in G. This can most conveniently be seen by rewriting the objective function as summation over the edges in G:

$$x^T A x = \sum_{(i,j) \in E} 2 x_i x_j$$

The factor of 2 is due to the symmetry of the adjacency matrix. For notational convenience, in the remainder, we maintain the identification of the index set of x with the set $V = \{1, \ldots, d\}$ of nodes and all considered graphs G are meant to have this node set.

It follows directly from Theorem 4.1 that StQP is an NP-hard problem.

Several authors have studied StQP and proposed solution methods that are exploiting the relationship with the max-clique problem. Bomze [Bom98] coined the term and proposes a reformulation that ensure the equality constraint by an appropriate objective penalty. [BLT08] reviews and compares several bounds on the problem. [Bom+00] uses the above mentioned reformulation as copositive programming problem and employs an interior-point method. [ST08] proposes a combinatorial enumeration algorithm for the problem. Approximation results for standard quadratic programming are found in [Nes99; BD02], where [BD02] derives a polynomial-time approximation scheme (PTAS) using appropriate SDP relaxations.

Our goal is to exploit Theorem 4.1 to obtain strong convex relaxations of StQP for general Q in the context of the Q-space relaxation and a spatial branch-and-bound approach.

4.2 Q-Space reformulation for StQP

The first step to solve StQP by spatial branch-and-bound approaches it to construct a convex relaxation of the problem that is then iteratively refined by branching.

Here, we place ourselves in the context of a reformulation of the problem where all the non-convex terms of the objective function are replaced with linearization variables, i.e., the Q-space reformulation from Section 3.2.2. We say that a term $Q_{ij}x_ix_j$ is convex if and only if $i = j$ and $Q_{ij} \geq 0$ or $i \neq j$ and $Q_{ij} = 0$. Accordingly, the objective matrix Q is decomposed into $Q = S + P$, where S contains all positive diagonal entries of Q and $P = Q - S$. Linearization variables Y_{ij} are introduced for all nonzero entries of P and $Y_{ij} = x_ix_j$ has to hold in every feasible solution. By abuse of notation, we interpret the linearization variables as a matrix Y for which the equation

$$Y = xx^T,$$

holds with the understanding that $Y_{ij} = x_ix_j$ has to hold whenever $P_{ij} \neq 0$. The reformulated StQP then reads as

$$\min \left\{ x^T S x + \langle P, Y \rangle \,\middle|\, Y = xx^T, x \in \Delta \right\}.$$

This formulation has a convex quadratic objective function and all the nonconvexities have been moved into the constraint $Y = xx^T$.

Once this reformulation is performed a convex relaxation can be formed by taking any convex relaxation of the feasible set

$$\Gamma = \left\{ (x, Y) \in \mathbb{R}^d \times (\mathbb{R}^d \times \mathbb{R}^d) \,\middle|\, Y = xx^T, x \in \Delta \right\}. \tag{4.1}$$

In our initial relaxation, we use the lower bounds $\underline{x}_i = 0$ and upper bounds $\overline{x}_i = 1$ and obtain the convex QP relaxation of StQP

$$\min \left\{ x^T S x + \langle P, Y \rangle \,\middle|\, \begin{matrix} \max\{0, x_i + x_j - 1\} \leq Y_{ij} \leq \min\{x_i, x_j\}, \\ x \in \Delta \end{matrix} \right\}. \tag{MC-StQP}$$

We refer to this relaxation as the Q-*space relaxation*.

While $Y = xx^T$ has to hold for each feasible solution, the McCormick inequalities only give a coarse approximation of the set Γ. We therefore strive to find valid inequalities that tighten this set and the Q-space relaxation. One very effective class of inequalities are the RLT inequalities from Section 3.3. Indeed, in the context of optimization over the standard simplex Δ, the only constraint that is not a bound constraint is $\sum_{i=1}^{d} x_i = 1$. In the first step of the RLT procedure, this equation is multiplied by one of the variables x_j, which yields the equation

$$\sum_{i=1}^{d} x_ix_j = x_j. \tag{4.2}$$

In the second step, the quadratic and bilinear terms are replaced with the linearization variables

$$\sum_{i=1}^{d} Y_{ij} = x_j. \tag{4.3}$$

As described in Section 3.3, linear over- and under-estimators can be used to project out terms that have not been linearize avoiding the introduction of new linearize variables. With $u_{ij}(x_i, x_j)$ and $o_{ij}(x_i, x_j)$ being linear under- and over-estimators, respectively, the projected RLT inequalities are:

$$\sum_{i \in V_j} Y_{ij} + \sum_{i \notin V_j} o_{ij}(x_i, x_j) \geq x_j, \tag{4.4}$$

$$\sum_{i \in V_j} Y_{ij} + \sum_{i \notin V_j} u_{ij}(x_i, x_j) \leq x_j. \tag{4.5}$$

In our implementation, we use

$$o_{ij}(x_i, x_j) = \overline{x_i} x_j$$
$$u_{ij}(x_i, x_j) = \underline{x_i} x_j.$$

For bilinear terms the McCormick over- and under-estimators would also be natural choices. In our experiments, the objective matrix is dense and thus only the diagonal entries of P might be zero.

A second stronger relaxation, denoted by RLT-stQP, is obtained by multiplying the standard simplex $\sum_{i=1}^{d} x_i = 1$ by each x_j, $j = 1, \ldots, d$ and adding the equation (4.3) or the inequalities (4.4) and (4.5) obtained.

Based on such a relaxation, a spatial branch-and-bound can then be performed by branching on the variables x and tightening the convex relaxation of Γ with the resulting local bounds at each node as described in Section 2.2.3.

4.3 Motzkin-Straus Clique inequalities

We now come back to Theorem 4.1 and its use. On the one side, Theorem 4.1 can be seen as a method to compute the clique number of a graph, but on the other side—as soon as the clique number of the graph is known—a valid inequality for Γ can be derived.

Corollary 4.2: *For any simple, undirected graph G with adjacency matrix A and clique number $\omega(G)$, the following inequality is valid for $(x, Y) \in \Gamma$:*

$$\langle A, Y \rangle \leq 1 - \frac{1}{\omega(G)}$$

Proof. The inequality $x^T A x \leq 1 - \frac{1}{\omega(G)}$ for $x \in \Delta$ follows immediately from the Motzkin-Straus theorem. Reformulation using the definition of Γ yields the result. □

In the remainder, we call the inequalities derived from Corollary 4.2 *Motzkin-Straus Clique inequalities (MSC inequalities)*.

For any instance of Γ one can derive a different MSC inequality from any graph G on d nodes. However, the following Theorem establishes that the inequalities stemming from certain graphs are dominated.

Theorem 4.3: *Let $G = (V, E)$ be a subgraph of $\tilde{G} = (V, \tilde{E})$ with the same clique number $\omega(G) = \omega(\tilde{G})$. Then, every point that violates the MSC inequality corresponding to G also violates the MSC inequality corresponding to $\tilde{G}$.*

Proof. Let the point (x, Y) be violated by the MSC inequality corresponding to G. Let A and $\tilde{A}$ be the adjacency matrices of G and $\tilde{G}$, respectively. Since G is a subgraph of $\tilde{G}$, it holds[1] $A \leq \tilde{A}$. Then, with $Y \geq 0$ we have

$$\langle \tilde{A}, Y \rangle \geq \langle A, Y \rangle > 1 - \frac{1}{\omega(G)}$$

Therefore, the MSC inequality corresponding to $\tilde{G}$ is also violated. □

We are therefore mostly interested in graphs that are "maximal" for a certain clique number in the sense that adding any edge increases their clique number.

If the original quadratic objective Q of StQP is the adjacency matrix of a graph, then the relaxation obtained by adding the corresponding MSC inequality to the Q-space relaxation of Γ has the same objective function of StQP. Because neither relaxations MC-StQP nor RLT-stQP solve general clique problems directly, the Motzkin-Straus Clique inequalities are not dominated by McCormick inequalities and RLT inequalities. This observation is indeed formalized in the following lemma, whose proof shows that a feasible point for the RTL-stQP relaxation can violate a Motzkin-Straus Clique inequality obtained from a graph G with clique number 2. Before proceeding to the statement of the lemma, we remind the reader that a complete bipartite graph with partition $(M, \bar{M})$ (with $M \subset V$) is a graph where every node in M is connected to all nodes in $\bar{M}$, but there are no edges between any pair of nodes in M or in $\bar{M}$. An obvious property of bipartite graphs is that their clique number is 2 and the MSC inequality corresponding to a bipartite graph is given by $\sum_{j \in M} \sum_{i \in \bar{M}} Y_{ij} \leq \frac{1}{4}$.

---

[1]Unless otherwise stated, we understand comparisons between two matrices and between a matrix and a scalar componentwise.

Theorem 4.4: *Let $G = (V, E)$ be a complete bipartite graph, the Motzkin-Straus Clique inequality obtained from G is not implied by RLT equation.*

Proof. We proof this by providing a point that is in RTL-stQP and that violates the MSC inequality corresponding to G. Let $(M, \bar{M})$ be the partition of V induced by G and $m = |M|$. Define a point $(\tilde{x}, \tilde{Y})$ as

$$\tilde{x}_i = \begin{cases} \frac{1}{2m} & \text{if } i \in M \\ \frac{1}{2(d-m)} & \text{otherwise} \end{cases}$$

$$\tilde{Y}_{i,j} = \begin{cases} \frac{1}{2m(d-m)} & \text{if } (i,j) \in (M \times \bar{M}) \cup (\bar{M} \times M) \\ 0 & \text{otherwise} \end{cases}$$

The notation $(i, j) \in (M \times \bar{M}) \cup (\bar{M} \times M)$ means that exactly one of the two indices is in M and the other in $\bar{M}$.

To show $\tilde{x} \in \Delta$, first observe that $\tilde{x} \geq 0$ and then we verify that $\tilde{x}$ satisfies the simplex constraint

$$\sum_{i=1}^{d} \tilde{x}_i = \sum_{i \in M} \frac{1}{2m} + \sum_{i \in \bar{M}} \frac{1}{2m(d-m)}$$
$$= \frac{m}{2m} + \frac{d-m}{2(d-m)}$$
$$= 1.$$

Next, we show that $(\tilde{x}, \tilde{Y})$ fulfills the McCormick inequalities. For the bounds $x \in [0, 1]^d$ these can be rewritten as

$$\max\{0, x_i + x_j - 1\} \leq Y_{ij} \leq \min\{x_i, x_j\}. \tag{4.6}$$

With the observation that $\tilde{x}_i \leq \frac{1}{2}$ for every i, we get that $\tilde{x}_i + \tilde{x}_j - 1 \leq 0$ and thus the first inequality becomes $0 \leq \tilde{Y}_{ij}$ which holds for all indices $i, j \in V$. The seconds inequality holds if $(i, j) \in (M \times \bar{M}) \cup (\bar{M} \times M)$ as $2m(d-m) \leq 2m$ and $2m(d-m) \leq 2(d-m)$ and thus, $\tilde{Y}_{ij} \leq \tilde{x}_i$ and $\tilde{Y}_{ij} \leq \tilde{x}_j$. Otherwise, clearly $0 \leq \tilde{x}_i$ for every index $i \in V$.

Furthermore, $(\tilde{x}, \tilde{Y})$ fulfills the RLT equations (4.3) for every $j \in V$. Indeed, for $j \in M$

$$\sum_{i \in V} \tilde{Y}_{ij} = \sum_{i \in \bar{M}} \tilde{Y}_{ij} = \frac{d-m}{2m(d-m)} = \frac{1}{2m} = \tilde{x}_j$$

and for $j \in \bar{M}$

$$\sum_{i \in V} \tilde{Y}_{ij} = \sum_{i \in M} \tilde{Y}_{ij} = \frac{m}{2m(d-m)} = \frac{1}{2(d-m)} = \tilde{x}_j.$$

As projecting out linearization variables by replacing them with proper estimators provides weaker inequalities, also all projected RLT inequalities are fulfilled.

However, $\tilde{Y}$ violates the Motzkin-Straus Clique inequality for the bipartite graph corresponding to the partition $(M, \bar{M})$

$$\sum_{j \in M} \sum_{i \in \bar{M}} \tilde{Y}_{ij} = \sum_{j \in M} \frac{d-m}{2m(d-m)} = \frac{m(d-m)}{2m(d-m)} = \frac{1}{2} > \frac{1}{4}.$$

$\square$

Theorem 4.4 again stresses the necessity to consider all problem constraints to construct tight convex relaxations for quadratic problems. Disregarding the simplex constraint $\sum_{i=1}^{d} x_i = 1$ and considering only the bounds constraints, the expression $\sum_{j \in M} \sum_{i \in \bar{M}} Y_{ij}$ is best possibly approximated by the termwise McCormick relaxation as it provides the convex full if the support graph is bipartite, see Corollary 3.2. Taking the simplex constraint into account, the MSC inequality provides a cut with a large violation and thus considerably strengthens the relaxation.

Even though the Motzkin-Straus Clique inequalities are not dominated by the RLT equations, a close relation exists. In particular, if $G = (V, E)$ is a complete graph, it is easy to see that aggregating the RLT constraints leads to an equality that dominates the corresponding Motzkin-Straus Clique inequality. Indeed, summing up the equations (4.3) for all $j \in V$ yields

$$\sum_{j \in V} \sum_{i \in V} Y_{ij} = \sum_{j \in V} x_j = 1$$

The last equation holds because x is in the standard simplex. Moving the quadratic terms to the right-hand side and observing that $\min_{x \in \Delta} \sum_{i \in V} x_i^2 = d^{-1}$, the Motzkin-Straus Clique inequality for the complete graph with d vertices can be derived

$$\sum_{j \in V} \sum_{\substack{i \in V \\ i \neq j}} Y_{ij} = 1 - \sum_{j \in V} x_j^2 \leq 1 - \frac{1}{d}$$

A more involved aggregation can be used to show the validity of Motzkin-Straus Clique inequalities for complete bipartite graphs. Consider a set $M \subset V$. First, sum up the equations (4.2) obtained by multiplying the standard simplex constraint for each x_j corresponding to $j \in M$, to obtain

$$\sum_{j \in M} \sum_{i \in V} x_i x_j = \sum_{j \in M} x_j.$$

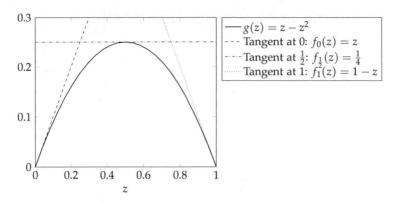

Figure 4.1: The function $g(z)$ with tangents at 0, $\frac{1}{2}$, and 1.

Next, regroup all the terms that have both variables in M in the right-hand side and obtain

$$\sum_{j\in M}\sum_{i\in \bar{M}} x_i x_j = \sum_{j\in M} x_j - \sum_{j\in M}\sum_{i\in M} x_i x_j$$

$$= \sum_{j\in M} x_j - \left(\sum_{j\in M} x_j\right)^2$$

Note that, so far, the linearization variables Y were not used and this last step used basic algebra to factor the right-hand side (this last step would not be satisfied by Y if only looking at the RLT inequalities). In the next step, we linearize the products on the left-hand side using Y:

$$\sum_{j\in M}\sum_{i\in \bar{M}} Y_{ij} = \sum_{j\in M} x_j - \left(\sum_{j\in M} x_j\right)^2. \tag{4.7}$$

Now, the right-hand side is the function $g(z) = z - z^2$ applied to $\sum_{j\in M} x_j$. Basic calculus tells us that $g(z)$ attains its maximum at $g(\frac{1}{2}) = \frac{1}{4}$ (see Fig. 4.1 for a plot of $g(z)$ on the domain of interest $[0,1]$). Therefore, we get that the right-hand side is smaller than or equal to $\frac{1}{4}$, which is exactly the Motzkin-Straus Clique inequality for the complete bipartite graph with partition $(M, \bar{M})$.

4.4 Generalized MSC inequalities for bipartite graphs

In the previous section, we have introduced Motzkin-Straus Clique inequalities and showed how, if G is a complete bipartite graph, the corresponding Motzkin-Straus Clique inequality can also be deduced by performing a specific aggregation of RLT inequalities. In this section, we generalize this reasoning and deduce a new class of cutting planes that can be obtained from bipartite graphs.

Note that to go from (4.7) to the Motzkin-Straus Clique inequality we used a constant over-estimator of g but, due to the concavity of g, every tangent overestimates g so that for the tangent f_α taken at α, the following inequality holds:

$$\sum_{j \in M} \sum_{i \in M} Y_{ij} \leq f_\alpha \left(\sum_{j \in M} x_j \right). \tag{4.8}$$

Because $f_\alpha(z)$ is an affine function, the right hand side of (4.8) is linear in x.

Of course, any missing linearization variable Y_{ij} can also be projected out in a similar way as in the RLT case. This way, the inequality using a tangent f_α becomes

$$\sum_{j \in M} \sum_{i \in M \cap V_j} Y_{ij} + \sum_{j \in M} \sum_{i \in M \cap \bar{V}_j} u_{ij}(x_i, x_j) \leq f_\alpha \left(\sum_{j \in M} x_j \right). \tag{4.9}$$

We denote constraint (4.9) as *generalized MSC bipartite inequality (GMSC bipartite inequality)*, and that depends on the choice of the point α where the tangent is taken and of the subset M. It turns out that, regardless of the choice of the partition $(M, \overline{M})$, the tangent obtained from $\alpha = 0$ and $\alpha = 1$ are always implied by the RLT inequalities (4.5).

Theorem 4.5: *If a point $(x, Y) \geq 0$ satisfies the RLT inequality (4.5) for all $j \in V$, then it satisfies the generalized MSC bipartite inequalities (4.9) for $\alpha = 0$ and $\alpha = 1$ and for all $M \subset V$.*

Proof. Take any $M \subset V$. We assume without loss of generality that the under-estimators $u_{ij}(x_i, x_j)$ are chosen non-negative. Since $(x, Y) \geq 0$ and because of the validity of the RLT inequalities (4.5), the following chain of inequalities is valid for every $j \in V$:

$$\sum_{i \in \overline{M} \cap V_j} Y_{ij} + \sum_{i \in \overline{M} \cap \bar{V}_j} u_{ij}(x_i, x_j) \leq \sum_{i \in V_j} Y_{ij} + \sum_{i \in \bar{V}_j} u_{ij}(x_i, x_j) \leq x_j.$$

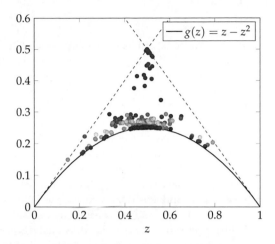

Figure 4.2: Violated points separated by generalized MSC bipartite inequalities. Notice that each point corresponds to a different set M.

Summing over $j \in M$, we get

$$\sum_{j \in M} \sum_{i \in \bar{M} \cap V_j} Y_{ij} + \sum_{j \in M} \sum_{i \in \bar{M} \cap \bar{V}_j} u_{ij}(x_i, x_j) \le \sum_{j \in M} x_j = f_0\left(\sum_{j \in M} x_j\right), \qquad (4.10)$$

which is the generalized MSC bipartite inequality for M at $f_0(z) = z$.

For the generalized MSC bipartite inequality at $f_1(z) = 1 - z$, it suffices to exchange M and $\bar{M}$ in (4.10) and, due to $x \in \Delta$, it holds that

$$\sum_{j \in M} x_j = 1 - \sum_{j \in \bar{M}} x_j.$$

$\square$

Figure 4.2 illustrates the generalized MSC bipartite inequalities than are separated in addition to RLT inequalities for one specific instance. More precisely, we separate generalized MSC bipartite inequalities as long as they are violated by using the separation algorithms that will be discussed in the next section. Each point in Fig. 4.2 has the value $z = \sum_{i \in M} x_i^*$ on x-axis and the value $\sum_{i \in M} \sum_{j \in \bar{M}} Y_{ij}^*$ on the y-axis and corresponds to a different relaxation solution (x^*, Y^*) and a different set M. The color indicates the round in which the point was separated and warmer colors mean that it was found later in the cut loop. The plot clearly illustrates that the tangents at 0 and 1 are implied for sets M by the RLT inequalities, but many additional cutting planes can be separated.

4.5 Separation

To separate a violated Motzkin-Straus Clique inequality, a graph has to be determined and its clique number has to be computed. On top of the fact that the latter computation is NP-hard, this boils down to a *bilevel separation problem* [LRW14] with the determination of the graph in the first level and the computation of the clique number in the second one. This turns out to computationally very hard because bilevel (integer-integer) optimization problems are, in general, extremely challenging both in theory and in practice (see, e.g., [Cap+16]).

The next sections describe two approaches to circumvent these difficulties. In Section 4.5.1 we focus on finding graphs with fixed clique number, especially bipartite graphs, and devise exact separation algorithms. That corresponds to removing the second level and the task is to find an appropriate graph. In Section 4.5.2 we use the solution of the continuous relaxation as starting point to heuristically find graphs that yield violated inequalities. This approach corresponds to eliminating the first level. It turns out that computing the clique number of instances of relevant sizes is still computationally feasible.

4.5.1 Exact separation

In the following, we will study mathematical programming formulations to separate MSC inequalities on specific classes of graphs striving for exact separation. This is computationally viable by restricting ourselves to graphs with known clique number so as to avoid solving bilevel programming problems (to separate a single cut). First, we will concentrate on general graphs with fixed clique number. Iterating over all possible clique sizes yields a separation algorithm for general graphs. Second, we focus on complete bipartite graphs, which clearly have clique number 2.

Graphs with fixed clique number. Consider a point (x^*, Y^*) and a fixed integer $k > 1$. The aim is finding a graph with clique size at most k whose corresponding Motzkin-Straus Clique inequality separates (x^*, Y^*) from Γ. Since $x^* \in \Delta$, without loss of generality (x^*, Y^*) can be assumed to be non-negative. Then, the following

Mixed-Integer Linear Programming problem (MILP) serves the purpose:

$$\max \quad \langle A, Y^* \rangle - \left(1 - \frac{1}{k}\right) \tag{4.11a}$$

$$\text{s.t.} \sum_{\substack{i,j \in S \\ i < j}} A_{ij} \le \frac{|S|(|S| - 1)}{2} - 1 \qquad \text{for all } S \subseteq V, |S| = k + 1 \tag{4.11b}$$

$$A_{ij} = A_{ji} \qquad\qquad \text{for all } i, j \in V \tag{4.11c}$$

$$A_{ij} = 0 \qquad\qquad \text{for all } i \notin V_j \tag{4.11d}$$

$$A_{ij} \in \{0, 1\} \qquad\qquad \text{for all } i, j \in V \tag{4.11e}$$

The program maximizes the violation of the cut and (x^*, Y^*) can be separated if and only if the objective value is greater than 0. Since the clique size is fixed, only the graph (in form of its adjacency matrix A) has to be computed. Constraints (4.11c)–(4.11e) ensure that A is indeed the adjacency matrix of a simple undirected graph. The inequalities (4.11b) ensure that G contains no clique of size $k + 1$. To this end, it requires that from every set of $S \subseteq V$ of cardinality $k + 1$, at least one if the $\frac{|S|(|S|-1)}{2}$ possible edges is missing.

A *posteriori*, if the clique size of G is smaller than k, then all edges connecting a maximum clique with the rest of the nodes have zero weight. Adding enough of these nodes will yield a graph with clique size k and the same objective value.

The drawback of this formulation is its exponential size for fixed k, which makes it impractical for computational purposes.

Complete Bipartite Graphs. We now turn to the separation of Motzkin-Straus Clique inequality stemming from complete bipartite graphs with partition $(M, \bar{M})$. We will always assume that both sets in the partition are nonempty and restrict ourselves to complete bipartite graphs (bipartite graphs such that adding any edge forms a triangle), since these are maximal w.r.t. the clique number and thus yield the strongest inequalities. For our purposes, bipartite graphs have two advantages: First, the clique number is 2 and therefore the Motzkin-Straus Clique inequalities have the best right-hand side value. Second, they have a very clean structure. For a fixed subset $M \subsetneq V$ of nodes, the Motzkin-Straus Clique inequality corresponding to the complete bipartite graph with partition $(M, \bar{M})$ is

$$\sum_{i \in M} \sum_{j \in \bar{M}} 2Y_{ij} \le 1 - \frac{1}{2}.$$

Separating a maximally violated Motzkin-Straus Clique inequality corresponding to some complete bipartite graph means finding a bipartite graph with maximum weight, where the weight for each edge (i, j) is given by Y_{ij}^*. This is equivalent to finding a maximum-weight cut and is thus NP-hard [Kar72]. However, since both, the number of nodes and the cardinality of the support (i.e., the nonzero values) of Y^*, are typically relatively small, it is computationally feasible to separate by solving a simple binary QP. To this end, we introduce a binary variable z_i for each $i \in V$ and say that nodes whose variables take the same value are in the same set of the partition. The problem to be solved is

$$\max \quad \sum_{i \in V} \sum_{j \in V} 2Y_{ij}^* z_i (1 - z_j) - \frac{1}{2} \tag{4.12a}$$

$$\text{s.t.} \quad z \in \{0, 1\}^{|V|} \tag{4.12b}$$

We assume without loss of generality that Y^* is symmetric. The product $z_i(1 - z_j)$ ensures that Y_{ij}^* is counted if and only if $z_i = 1$ and $z_j = 0$, i.e., i and j are in different sets. The objective function therefore maximizes the violation of the cut. Every solution with positive objective function value corresponds to a violated cut with partition $(M, \bar{M})$, where $M = \{i \in V \mid z_i = 1\}$. If the optimal objective value is non-positive, no violated cut in this class exists.

Note that Y^* is part of the input in (4.12) and thus we are facing a binary QP. Products of binary variables however are easily linearized by introducing a linearization variable and coupling it with the two original variables by the usual McCormick inequalities. As the McCormick inequalities are exact of one of the variables is at their bounds, this formulation suffices and there is no need for spatial branching. Solvers, CPLEX in particular, will do this transformation automatically in preprocessing and solve (4.12) as an MILP.

Generalized MSC bipartite inequalities. To separate a violated generalized MSC bipartite inequality, a set M has to be found such that

$$\sum_{i \in M} \sum_{j \in M} Y_{ij}^* > g \left(\sum_{i \in M} x_i^* \right).$$

where g is still the function $g(z) = z - z^2$. The generalized MSC bipartite inequality for M and $\alpha = \sum_{i \in M} x_i^*$ will then separate this point. The separating binary QP for bipartite graphs (4.12) can be generalized to separate violated generalized MSC bipartite inequalities. It maximizes the violation and adds the α

as one of the decision variables. Namely,

$$\max \quad \sum_{i \in V} \sum_{j \in V} Y_{ij}^* z_i (1 - z_j) - (\alpha - \alpha^2) \tag{4.13a}$$

$$\text{s.t.} \quad \alpha = \sum_{i \in V} x_i^* z_i \tag{4.13b}$$

$$z \in \{0, 1\}^{|V|}, \alpha \geq 0. \tag{4.13c}$$

As for bipartite graphs, nodes are partitioned according to the value of their associated variables.

4.5.2 Heuristic separations and strengthening

Practically relevant exact separation methods for MSC inequalities are only available for bipartite graphs. Heuristic methods can be used to find more general graphs that yield violated MSC inequalities. Heuristic methods have a long tradition in mathematical programming [Ber14]. In contrast to exact methods, heuristics generally do not provide a guarantee of success, which is typically compensated by short(er) running times.

As discussed, a MSC inequality can be associated with any graph but, due to the computational difficulty of solving (4.11a)–(4.11e), trying promising candidates is a sensible approach. In the light of the bilevel separation problem this corresponds to fixing the first level of finding an appropriate graph and leaves only the computation of the clique number. Even though NP-hard in theory, computing the clique number for graphs of relevant sizes is computationally feasible for our application.

A very natural candidate to consider is the support graph of the current relaxation solution Y^*. This graph has an edge (i, j) if and only if $Y_{ij}^* \neq 0$. The drawback of this choice is that a lot of edges (i, j) with small weight Y_{ij}^* might be in the graph and increase its clique number (and thus weaken the right hand side of the MSC inequality) without bringing much benefit. This is addressed by a procedure to be introduce in the next paragraph.

Clique Reduction. The right-hand side of a Motzkin-Straus Clique inequality gets smaller with decreasing the clique number. Removing one edge from each maximum clique reduces the clique number by one and yields an inequality with stronger right-hand side. This can be used to generate more valid inequalities from a graph with high clique number and also to find a violated cut, e.g., if the originally considered graph does not correspond to a violated cut.

It is sensible in practice to remove the edge with smallest weight from every maximum clique. However, the following MILP determines an optimal subset of edges to be removed from a graph $G = (V, E)$:

$$\min \quad \sum_{(i,j) \in E} Y^*_{ij} z_{ij} \tag{4.14}$$

$$\text{s.t.} \sum_{\substack{i,j \in S \\ i<j}} z_{ij} \geq 1 \qquad \text{for each maximum clique } S \text{ in } G \tag{4.15}$$

$$z_{ij} \in \{0, 1\} \qquad (i, j) \in E \tag{4.16}$$

A variable z_{ij} takes the value 1 if the edge (i, j) should be removed. The objective function ensures that minimal weight is removed from the cut. The constraint (4.15) ensures that at least one edge from each maximum clique is removed. Of course, MILP (4.14)–(4.16) has exponentially many constraints in general, so it has to be managed with care.

Lifting. As seen in Theorem 4.3, a MSC inequality corresponding to a subgraph $\tilde{G}$ of a graph G with the same clique number is dominated by the MSC inequality corresponding to G. Therefore, a cut can be lifted by adding edges to the graph as long as the clique number does not increase. Of course, complete bipartite graphs are maximal in this respect but other, possibly heuristically generated, graphs might not be.

The implementation details of the heuristic separation methods are presented in Section 4.6.4.

4.6 Computational Experiments

In this section, we present the results of a large set of computational experiments. They were carried out on a cluster of Intel Xeon 5160 quadcore CPUs running at 3.00 GHz with 8 GB RAM and using RHEL5 as operating system. To avoid random noise by cache misses and alike only one process was executed on each node at a time.

The implementation is based on a slightly modified version of the IBM CPLEX Optimizer 12.6.3 [IBMb] (CPLEX for short) where the C-API has been extended to provide callbacks the access to the linearization variables Y. The cuts are separated from the user cut callback, only at the root node, and are added with the *purgeable flag* set to CPX_USECUT_PURGE, in order to allow CPLEX to purge the cuts that are deemed to be ineffective according to its internal strategies.

Our computational investigation focuses on the application of the proposed cutting planes in a Spatial branch-and-bound algorithm and studies their impact on the root node and on the overall solution time. Therefore, we omit a direct comparison with alternative formulations or solution approaches as presented for example in [ST08]. Nevertheless, in Section 4.6.6 we report computational results on the small set of publicly-available instances from [ST08], while in the next section we describe the large amount of randomly-generated instances we extensively based our computation on.

This section is organized as follows: In Section 4.6.1, we describe the instances that were used in the computational study. Section 4.6.2 compares the closures of the different classes of inequalities and analyzes differences between MSC bipartite inequalities and GMSC bipartite inequalities. To this end, only the root node of the branch-and-cut tree is considered. In Section 4.6.3, hybrid configurations that combine the advantages of the different classes of inequalities are introduces and evaluated. While all these sections focused on cuts from bipartite graphs, Section 4.6.4 focuses on the heuristic separation of Motzkin-Straus Clique inequalities from general graphs, i.e., also from graphs with a clique number greater than 2. After this extensive study on the impact on the root node, Section 4.6.5 summarized results on the overall solution process. Finally, results on a small testset from the literature are presented in Section 4.6.6.

In this section, we present aggregated results in several tables for various different configurations. Detailed per-instance tables for all configurations can be found in the appendix of the online version of this thesis [Sch17].

4.6.1 Instances

As anticipated, we considered a large set of randomly-generated instances, and, in particular, we considered two sizes, $d = 30$ and $d = 50$, so as only the objective matrix has to be sampled. The instances are available on `http://or.dei.unibo.it/library/msc`

The sign of the objective coefficients plays a major role in these instances. Assuming all terms are linearized (i.e., all diagonal entries are negative and all off-diagonal entries are not zero), the objective function only acts on the Y variables. When optimizing the value of Y over Γ, Y_{ij} with $i \neq j$ is restricted by the McCormick inequalities and whether it will take the upper or the lower bound is defined by the sign of Q_{ij}, namely

$$Y_{ij} = \begin{cases} \max\{0, x_i + x_j - 1\} & \text{if } Q_{ij} > 0, \\ \min\{x_i, x_j\} & \text{if } Q_{ij} < 0. \end{cases}$$

We therefore strive to generate instances with different fractions of positive and negative entries in Q. Since the inequalities presented in this paper can only cut points where at least some entries Y_{ij} exceed $x_i x_j$, the biggest impact is expected for instances with a lot of negative entries in Q.

We used triangular distributions, which are characterized by 3 parameters $a < c < b$. Namely, a and b are the minimum and the maximum of the value range. The mode c describes the peak of the piecewise linear density function. Of course, the sign of the coefficients is of great impact in general; on the diagonal they even decide if the respective terms are convex. We therefore use the triples $(-10, -5, 0)$ and $(0, 5, 10)$ to get instances with only negative and only positive coefficients. For instances with mixed signs, we used the triples $(-10, -3, 10)$, $(-10, 0, 10)$, and $(-10, 3, 10)$, where the second is a symmetric distribution and the other two are more likely to have positive or negative coefficients, respectively. The diagonal entries are divided by 2. In addition, 2 variants of each instance are generated by taking the positive and negative absolute values of the diagonal elements. For $(-10, -5, 0)$ only the positive and for $(0, 5, 10)$ only the negative variants are generated since the respective opposite would yield the same instance again. Furthermore, the instances with positive off-diagonal entries (distribution $(0, 5, 10)$) in the variant with negative diagonal entries are trivially solved by all approaches. Indeed, since the objective is to minimize, setting the variable x_i with lowest diagonal entry Q_{ii} in the objective to 1 gives the optimal solution. For this reason, those instances are excluded.

The 3 distributions in 3 variants, 1 distributions in 2 variants for the diagonal, and 1 distribution in 1 variant for the diagonal give 12 different instance types. For each instance type and for each size $d \in \{30, 50\}$ we generated 10 instances, yielding 120 instances with $d = 30$ and 120 instances with $d = 50$ in total. All instances are available on http://or.dei.unibo.it/library/msc. In all computational experiments we enforced a time limit of 2 hours for instances of size $d = 30$ and 6 hours for those of size $d = 50$.

Since we can only separate MSC and GMSC bipartite inequalities if the linearization variable Y_{ij} exceeds the respective product, i.e., $Y_{ij} > x_i x_j$, for some (i, j), one could assume that the instances $(0, 5, 10)$ will not be affected by Motzkin-Straus Clique inequalities given that the objective function drives the linearization variables towards 0. This is only true for instances with positive diagonal elements in Q. For these instances, the quadratic terms $Q_{ii} x_i^2$ are convex and thus not linearized, and the resulting projected RLT inequalities are redundant. For the variation with negative diagonal elements in Q, the quadratic terms are linearized and the RLT equations $\sum_i Y_{ij} = x_j$ for all $j \in V$ force some Y_{ij} to be positive. In all these instances it is then possible to separate MSC inequalities and generalized

MSC bipartite inequalities.

4.6.2 Optimizing over the bipartite closures

As a first set of experiments, we want to evaluate the impact of Motzkin-Straus Clique inequalities corresponding to bipartite graphs and generalized MSC bipartite inequalities at the root node of the branch-and-cut tree. Since RLT equations and inequalities can be easily separated by enumeration and are expected to be effective, we separate our inequalities only if no violated RLT inequalities can be found.

For MSC and GMSC bipartite inequalities we have exact separation algorithms and thus we can optimize over the associated closures. The closure of a class of inequalities is the relaxation obtained by adding all possible inequalities of this class. Although the closure itself may be intractable to compute, one can optimize a linear function over it by separation. The comparison of the values gives an indication of the strength of the class of inequalities.

MSC and GMSC bipartite inequalities are separated by solving the associated mathematical models (4.12) and (4.13) in a classical cutting-plane scheme, by using CPLEX as a black box. To limit the tailing-off effect that often arises in cutting plane algorithms, we try to separate up to 5 cuts per round, i.e., at every separation round we collect the first 5 incumbent solutions returned by CPLEX that correspond to violated cuts. Specifically, we consider the following four settings:

CPLEX   CPLEX with empty cut callback;

RLT   Violated RLT equations and inequalities are added;

Bipartite   At each call of the cut callback, first violated RLT equations and inequalities and then violated MSC bipartite inequalities are added. If no inequality can be separated anymore, the final dual bound gives the value of the closure over these two types of cuts;

GMSC   Same as Bipartite, but GMSC bipartite inequalities are separated instead of MSC bipartite inequalities.

Each of these configurations improves the closure of the previous ones since Bipartite and GMSC also separate RLT equations and inequalities and since Motzkin-Straus Clique inequalities for bipartite graphs are generalized MSC bipartite inequality at $\alpha = 0.5$.

Tables 4.1 and 4.2 report aggregated results at the root node for these configurations on the instances of size $d = 30$ and $d = 50$, respectively. Both tables have the following structure: First, we report the average root gap to measure the strength of the separated cuts. For all considered approaches we give the %gap left at the root, computed as $(UB - LB_{root})/|UB|$, where LB_{root} is the dual bound at the root node and UB is the optimal solution value or the value of the best solution found by any of the approaches reported in Section 4.6.5. Then we report the gap closed with respect to CPLEX root (resp. CPLEX with RLT inequalities), computed as $(LB_{root} - LB_{base})/(UB - LB_{base})$, where LB_{base} is the dual bound obtained by CPLEX root (resp. CPLEX with RLT). Then, the number of instances solved to proven optimality is given, along with the number of time limits hit. Next, we give the average and maximum separation time, first considering all instances and then disregarding the instances where any of the compared approaches hit the time limit. In addition, we report the average and maximum number of separated cuts, along with the number of cuts applied to the root LP at the end of root node, as reported by CPLEX. Finally, the average and maximum number of separation rounds is given, to specify how many times the callback was called (note that in the last round no cut was separated otherwise the callback would have been called again). In both tables the two last columns, GraphPool and Hybrid, correspond to configurations being introduced in Section 4.6.3.

The results reported in the tables clearly show that RLT inequalities are fundamental for StQP. Indeed, by themselves RLT inequalities already close about 85 % of the root gap obtained by default CPLEX. On the other hand, MSC and GMSC bipartite inequalities are very effective to improve on the dual bound on top of RLT inequalities, and the GMSC bipartite closure appears definitely stronger than the MSC bipartite closure. Bipartite and GMSC greatly improve over RLT reducing the arithmetic mean of the root gap and GMSC gives the best dual bounds by a large amount. Concerning the separation time, Bipartite appears on average to be very fast, while GMSC is instead too time consuming. With GMSC, 3 instances of size $d = 30$ and 10 of size $d = 50$ do not finish the root node within the time limit. However, it is remarkable that GMSC is able to solve 3 instances of size $d = 30$ without resorting to branching.

To investigate the main differences between Bipartite and GMSC we looked closely at the evolution of the root node for some specific instances. All plots reported in the following are given on one instance of size $d = 30$ with positive diagonal entries and distribution $(-10, 0, -5)$ (i.e., instance *triangular_30_-10_0_-5__04_posDiag*). The instance has been selected as the one on which the separation time of both Bipartite and GMSC exceeds the respective arithmetic mean by the smallest amount, but the plots would look similar for all instances.

Figure 4.3 plots the evolution of the dual bound from round to round (Fig. 4.3a) and w.r.t. the accumulated separation time (Fig. 4.3b). Even if GMSC converges

	CPLEX	RLT	Bipartite	GMSC
Average root gap [%]				
Gap left	1019.33	67.11	21.55	11.04
Closed wrt CPLEX root	–	84.95	90.02	91.22
Closed wrt RLT	–	–	68.81	86.19
Solved/Timeout at the root				
Solved	0	0	0	3
Timeout	0	0	0	3
Separation time in seconds				
Mean	0.00	0.00	1.24	425.39
Max	0.00	0.00	14.50	7139.10
Separation time in seconds (exclude timelimit)				
Mean	0.00	0.00	1.02	255.04
Max	0.00	0.00	14.60	5651.30
Number of cuts				
Separated Mean	0.00	26.98	97.62	748.83
Separated Max	0.00	30.00	382.00	4651.00
Applied Mean	0.00	26.98	53.72	84.67
Applied Max	0.00	30.00	137.00	228.00
Number of separation rounds				
Mean	0.00	1.92	47.29	225.28
Max	0.00	2.00	176.00	1032.00

Table 4.1: Comparing the closures of StQPs of size 30

towards a stronger dual bound, Bipartite is superior in the first rounds and shows a very limited tailing off effect. On the contrary, GMSC stalls after about 200 rounds and after that each round of cuts increases the bound only by a very small amount. Note that the differences would be even more marked by looking at the accumulated time spent. Indeed, Bipartite starts stalling after about 100 rounds and 7.3 seconds, with a dual bound of value -3.1285, and then it takes 17.2 seconds and 175 rounds to converge to the MSC bipartite closure of value -3.1088. GMSC needs less rounds but twice the time to reach the same dual bound of value -3.1271 where Bipartite started stalling, namely 88 rounds and 14.0 seconds. Stalling of Bipartite explains why less time and rounds to reach a dual bound of value -3.1087 that is similar to the MSC bipartite closure, namely

	CPLEX	RLT	Bipartite	GMSC
Average root gap [%]				
Gap left	1640.78	61.14	20.40	10.37
Closed wrt CPLEX root	–	87.94	90.73	91.46
Closed wrt RLT	–	–	68.45	86.79
Solved/Timeout at the root				
Solved	0	0	0	0
Timeout	0	0	0	10
Separation time in seconds				
Mean	0.00	0.00	18.07	2183.95
Max	0.00	0.00	533.10	21488.30
Separation time in seconds (exclude timelimit)				
Mean	0.00	0.00	2.23	439.81
Max	0.00	0.00	7.00	8266.00
Number of cuts				
Separated Mean	0.00	45.63	240.19	1771.27
Separated Max	0.00	50.00	2068.00	9230.00
Applied Mean	0.00	45.63	97.75	159.83
Applied Max	0.00	50.00	326.00	354.00
Number of separation rounds				
Mean	0.00	1.96	102.91	477.81
Max	0.00	3.00	583.00	2175.00

Table 4.2: Comparing the closures of StQPs of size 50

18.4 seconds and 98 rounds. Finally, GMSC saturates after about 200 rounds and 153.7 seconds with a dual bound of value -3.0299, and it converges to the GMSC bipartite closure of value -2.9943 in 2756.6 seconds and 717 rounds.

One difference between the two plots, Figs. 4.3a and 4.3b, is the extremely steep slope in the beginning of Fig. 4.3b. This indicates that the first rounds of separation are extremely fast and especially separation of generalized MSC bipartite inequalities slows down as it gets more difficult to find violated cuts. Figure 4.4 confirms this by plotting the time used in each round of separation for Bipartite and GMSC on the same instance. For Bipartite the separation times remain almost constant at a very low value. For GMSC, in contrast, separation times are modest for the first rounds but start to increase soon, with outliers taking up to more than 20 seconds. Such a difference in the separation time between `Bipartite` and `GMSC` can be easily explained: the former separation problem is a binary QP that can be linearized and solved by MILP techniques, while the latter has a non-convex quadratic continuous variable (namely α) and requires Spatial branch-and-bound to be solved.

Finally, we analyze the diversity of the graphs that are generated by `Bipartite` and `GMSC`. To this end, we compare each graph that is returned by the separation problems to all graphs that have been separated previously. The difference between two bipartite graphs is defined in terms of the partitions: Let $\mathcal{M} = (M, \bar{M})$ and $\mathcal{N} = (N, \bar{N})$ be two partitions of the same set. Then define the distance $d(\mathcal{M}, \mathcal{N})$ between the partitions by

$$d(\mathcal{M}, \mathcal{N}) = \min(|M \triangle N|, |M \triangle \bar{N}|)$$

where is the $\triangle$ is the symmetric difference. Note that $M \triangle N = \bar{M} \triangle \bar{N}$, so the above is well defined.

Figures 4.5 and 4.6 plots the minimum distance of every graph to all previous graphs for `Bipartite` (Fig. 4.5a) and `GMSC` (Figs. 4.5b and 4.6). Specifically, for each round, the picture reports the minimum distance of every graph generated at the given round. Since `Bipartite` requires only 175 rounds to converge against the 717 required by `GMSC`, the plots in Fig. 4.5 for `GMSC` is restricted to the first 175 rounds while Fig. 4.6 shows the same plot for `GMSC` and all rounds. The picture clearly shows that `Bipartite` tends to separate cuts associated with bipartite graphs that are more diverse with respect to the cuts separated by `GMSC`. While for `Bipartite` the vast majority of the graphs has a distance between 5 and 10 to the previously separated graphs, we see a lot of graphs that are very similar, e.g., distance smaller or equal to 2, to one of the previous graphs for `GMSC`. Frequently we even separate the same graph multiple times with different values of α. This is problematic since the resulting cuts in this case are very similar. Indeed, from more diversity in the observed graphs more diverse cuts emerge.

4.6.3 Combining the separation of MSC and GMSC bipartite inequalities

The results reported in the previous section have shown that MSC and GMSC bipartite inequalities are very effective to improve on the root node dual bound on top of RLT inequalities. Further, the GMSC bipartite closure appears definitely stronger than the MSC bipartite closure. However, while optimizing over the MSC bipartite closure is computationally feasible, separating GMSC bipartite inequalities can be rather expensive in practice and, in particular, optimizing over the GMSC bipartite closure by repeatedly solving the separation problem (4.13) is on average too time consuming.

In this section, we present some simple ideas to combine the separation of both types of inequalities, in order to overcome the main drawbacks of the GMSC approach discussed in Section 4.6.2, while trying to approximate the GMSC bipartite closure as much as possible.

Separating GMSC bipartite inequalities from the graph pool

A heuristic approach to separate a GMSC bipartite inequality without solving the corresponding separation problem (4.13) would be to first obtain a violated MSC bipartite inequality and the corresponding partition $(M, \bar{M})$ by solving the easier separation problem (4.12). Then, instead of adding the separated inequality, one could compute $\alpha = \sum_{i \in M} x_i^*$ a posteriori from the current relaxation solution (x^*, Y^*) and then add the GMSC bipartite inequality at α. The dual bound that can be obtained by this approach is in between the values of the two closures.

The idea we adopted follows this spirit even more radically. While separating MSC bipartite inequalities, we store the new graphs (i.e., the partitions) that get separated in a *graph pool*. After adding the MSC bipartite inequalities, we compute the GMSC bipartite inequalities from all graphs in the pool with respect to the current relaxation solution (x^*, Y^*) and add the violated ones. This approach is expected to generate a lot of very similar cuts at every round, and thus we rely on CPLEX cut purging to discard the ones that are not useful. Indeed, the experiments show that the number of cuts that are actually applied to the LP relaxation at the end of the root node is comparable to the number obtained with GMSC. The repeated separation of GMSC bipartite inequalities from the same graphs can be seen as a way to "update" the existing cuts based on the current relaxation solution to cut-off. Using the separator (4.12) from Bipartite ensures fast separation times and adding also the MSC bipartite inequalities ensures that the same graph cannot occur twice. In the following, we refer to this approach as GraphPool.

Aggregated results obtained at the root node with GraphPool approach on instances of size $d = 30$ and $d = 50$ are reported in the third column in Tables 4.3 and 4.4, respectively. The tables have the same structure as Table 4.1 and show that GraphPool is a reasonable trade-off between Bipartite and GMSC. On the one side, the average separation time is only a small fraction of the time needed by GMSC, and the outliers are under control, as no instance hits the time limit and the maximum separation time excluding the time limits is reduced from 5651.3 to 98.3 seconds for $d = 30$ and from 8266.0 to 230.2 seconds for $d = 50$.

	Bipartite	GMSC	GraphPool	Hybrid	HybridGP
Average root gap [%]					
Gap left	21.55	11.04	13.73	11.33	11.06
Closed wrt CPLEX root	90.02	91.22	90.94	91.20	91.22
Closed wrt RLT	68.81	86.19	81.93	85.69	86.17
Solved/Timeout at the root					
Solved	0	3	0	3	3
Timeout	0	3	0	0	0
Separation time in seconds					
Mean	1.24	425.39	7.23	17.07	48.61
Max	14.50	7139.10	118.30	142.00	799.30
Separation time in seconds (exclude timelimit)					
Mean	1.02	255.04	5.15	15.06	34.34
Max	14.60	5651.30	98.30	147.20	730.40
Number of cuts					
Separated Mean	97.62	748.83	2456.58	433.13	5565.64
Separated Max	382.00	4651.00	16491.00	782.00	32271.00
Applied Mean	53.72	84.67	78.16	77.34	104.12
Applied Max	137.00	228.00	200.00	168.00	256.00
Number of separation rounds					
Mean	47.29	225.28	96.39	216.73	161.24
Max	176.00	1032.00	515.00	376.00	710.00

Table 4.3: Comparing the "advanced" configurations on StQPs of size 30

On the other hand, separating GMSC bipartite inequalities from the graph pool yields a substantial reduction of the gap left at the root w.r.t. Bipartite. One final information it is worthy to remark is about the number of separated cuts. As already mentioned, GraphPool tends to generate a lot of very similar cuts at every round, and the average number of separated cuts dramatically increases w.r.t. both Bipartite and GMSC. However, most of the separated cuts are purged by CPLEX and the number of *applied* cuts at the end of the root node is similar to the one obtained with GMSC.

Figure 4.7 shows the number of cuts that get separated every round. After an initial ramp-up phase where almost all generalized MSC bipartite inequalities are updated, the number of updates diversifies with a tendency to decrease.

It is remarkable, that GraphPool dramatically increases the number of separation rounds until convergence compared to Bipartite. While Bipartite has

	Bipartite	GMSC	GraphPool	Hybrid	HybridGP
Average root gap [%]					
Gap left	20.40	10.37	13.66	11.20	10.44
Closed wrt CPLEX root	90.73	91.46	91.23	91.42	91.46
Closed wrt RLT	68.45	86.79	81.10	85.24	86.67
Solved/Timeout at the root					
Solved	0	0	0	0	0
Timeout	0	10	0	0	0
Separation time in seconds					
Mean	18.07	2183.95	203.22	92.98	355.65
Max	533.10	21488.30	4732.70	2097.00	5488.00
Separation time in seconds (exclude timelimit)					
Mean	2.23	439.81	20.02	28.17	64.34
Max	7.00	8266.00	230.20	121.40	352.70
Number of cuts					
Separated Mean	240.19	1771.27	12281.52	676.42	19473.42
Separated Max	2068.00	9230.00	76722.00	2634.00	62392.00
Applied Mean	97.75	159.83	150.72	136.15	194.48
Applied Max	326.00	354.00	447.00	349.00	479.00
Number of separation rounds					
Mean	102.91	477.81	285.40	285.59	356.26
Max	583.00	2175.00	1210.00	783.00	719.00

Table 4.4: Comparing the "advanced" configurations on StQPs of size 50

converged after 47 rounds on the testset of size 30 (103 rounds on the testset of size 50), GraphPool 96 (285) rounds on average. GMSC needs 225 (478) rounds on average to convergence, but also spends much more time per round.

Even if GraphPool closes a large part of the gap between Bipartite and GMSC, in order to better approximate the GMSC bipartite closure we need to directly separate GMSC bipartite inequality by solving the corresponding separation problem (4.13). However, as shown in Section 4.6.2, solving (4.13) can often be too time consuming. We tried to overcome this drawback in two ways.

Applying work limits in the separation of GMSC bipartite inequalities

One reason for the large separation times observed in GMSC is that, at each round, we try to separate up to 5 cuts instead of aborting the separation after finding the first violated cut. While the experiments show that this is indeed advantageous, the work to generate additional cuts should be limited to remove outliers. To maintain determinism, we apply a work limit in terms of deterministic ticks as supplied by the CPLEX API. In each separation round, we let τ_0 be the ticks needed to find the first violated cut and $\Phi > 0$ a scaling parameter. The additional work to find the ith violated cut after finding $i - 1$ is then limited by

$$\tilde{\tau}_i = \tau_0 \Phi^i$$

In our computations, we chose $\Phi = 0.9$, so allow less work on every iteration. The effect of this work limit has been tested on GMSC and it turned out to be quite beneficial. On the instances of size $d = 30$, only 1 instead of 3 instances does not finish the root node within the time limit of two hours. On those 117 instances where both approaches finished within the time limit, the average separation time decreases from 255 seconds to 215 seconds, even if the average number of separation rounds needed to convergence to the closure increases from 202 to 350. Last, the time for the longest separation round goes down from 98 to 66 seconds; an effect which is even more pronounced on the bigger test set with $d = 50$ where the time for the longest separation goes down from 12239 to 47 seconds. Indeed, on the instances with $d = 50$ it regularly happens that a small but not sufficient number of cuts is found in a reasonable time, but proving optimality or finding sufficiently many cuts then takes an outrageous amount of time.

Limiting the work to find additional cuts however does not solve the second problem GMSC: Tailing-off. To this end, we propose to limit the number of separation rounds. In order to keep things simple, we propose to apply a static limit on the number of iterations.

Experiments show that these measures already boost performance of GMSC, but the extremely fast running times of Bipartite are left unused. In the next section fast separation of bipartite Motzkin-Straus Clique inequalities and the strength of generalized MSC bipartite inequalities are combined.

Combining exact separation of MSC and GMSC bipartite inequalities

The comparison between Bipartite and GMSC has shown that MSC bipartite inequalities can be separated much more efficiently than GMSC bipartite inequalities. Also, as highlighted in Section 4.6.2, solving (4.12) instead of (4.13) favors the separation of more diverse cuts and this often leads to a larger improvement on the dual bound in the first rounds. To exploit the advantages of (4.12), we propose two hybrid variants where we first run one of Bipartite or GraphPool and then a limited number of rounds of GMSC, each round amended with the work limit previously described. Even tough applying GraphPool is typically fast for the smaller instances, some instances in the bigger test set of size $d = 50$ show a long tailing-off for GraphPool as well. We therefore add a limit in the number of rounds to protect against these outliers and test two hybrid settings:

Hybrid   Optimize over the MSC bipartite closure (i.e., perform Bipartite) and then separate up to 200 rounds of GMSC bipartite inequalities (i.e., perform up to 200 rounds of GMSC).

HybridGP   Separate MSC bipartite inequalities with (4.12) and GMSC bipartite inequalities from the graph pool (i.e., perform GraphPool) for up to 500 rounds and then perform up to 200 rounds of GMSC.

Aggregated results obtained at the root node with Hybrid and HybridGP are again reported in the last two columns in Tables 4.3 and 4.4. HybridGP gives a very tight approximation of the GMSC bipartite closure and it appears to be a clear improvement w.r.t. GMSC, as it allows to close almost the same %gap w.r.t. RLT inequalities in a fraction of the time required by GMSC, while there is no instance on which the time limit is hit. Hybrid yields a better dual bound w.r.t. GraphPool, it is considerably faster than HybridGP. However, it cannot close the same %gap as HybridGP, especially for the case $d = 50$ on which the differences between Hybrid and HybridGP are more marked.

4.6.4 Motzkin-Straus Clique inequalities for higher clique numbers

In this section we evaluate the impact of Motzkin-Straus Clique inequalities associated with general graphs with any clique number. To this end, we separate the inequalities by using the heuristics from Section 4.5.2, both as stand alone separation method and as a means to complement the exact separation of MSC bipartite inequalities discussed in the previous sections.

The heuristic is implemented as follows. The support graph $G = (V, A)$ of the relaxation solution Y^* is used a first candidate, i.e., an edge (i, j) is present in the graph if and only if $Y_{ij}^* > 0$. From this, by clique reduction a series of graphs with decreasing clique number is derived. More precisely, the edges (i, j) with lowest weight Y_{ij}^* is removed from every maximum clique. Each graph in the series is lifted by adding randomly selected edges that do not increase the clique numbers. We use a different random sequence for each of the graphs to make the cuts more diverse. All violated MSC inequalities for the lifted graphs are added to the problem. Clique numbers, maximum cliques and alike are computed by the igraph library [CN06].

Heuristic separation can be used in two ways: As stand-alone separation procedure and in combination with the very effective methods discussed in the previous sections. The configuration Heur only relied on heuristic separation. To study the interplay between heuristic and exact separation, the configurations `Bipartite` and `GraphPool` are chose as reference to be enhanced by heuristic separation. This is done in two ways: First, by applying the heuristic procedure in every round after the exact separator. This approach seems most promising as it is typically advantageous to separate several cutting planes in each separation round. Second, the heuristic is used to only in rounds, where the respective reference separator was not successful in finding a violated inequality. This approach gives an idea of how much the heuristic is able to improve the bound on top of the reference configuration. Overall, we use the following settings to assess the impact of the proposed heuristic separation methods:

Heur	Separate violated RLT equations and inequalities and heuristically separate Motzkin-Straus Clique inequalities as described above.
Bipartite→Heur	Perform `Bipartite`. Then separate cuts heuristically for one round and continue with `Bipartite`. Iterate at most 5 times.
Bipartite+Heur	While performing `Bipartite`, separate Motzkin-Straus Clique inequalities also heuristically in each round of separation.
GraphPool→Heur	Same as `Bipartite→Heur`, but with `GraphPool` instead of `Bipartite`.
GraphPool+Heur	Same as `Bipartite+Heur`, but with `GraphPool` instead of `Bipartite`.

Table 4.5 shows performance numbers for the root nodes and compares the heuristic settings to their reference configurations. To this end, average root gap, time needed for separation (i.e., the time that is effectively spent in the cut callback), and the overall root processing time is given. The table is split into two parts for instances of size $d = 30$ and $d = 50$ and in each part first the figures for Heur are given, followed by two blocks of the approaches that being derived from `Bipartite` and `GraphPool`, respectively. The average root gap achieved by Heur is comparable to the bipartite closure. Heuristic separation after `Bipartite` (i.e., `Bipartite→Heur`) only gives a very modest improvement of the dual bound. Simultaneous separation of MSC inequalities from bipartite graphs and general graphs, i.e., `Bipartite+Heur`, achieves a slightly better bound. Even though MSC inequalities from general graphs are separated via a heuristic and

	Avg. Root Gap		Sepa. Time [s]		Root Time [s]	
	Left	Cl. wrt RLT	Avg.	Max.	Avg.	Max.
Instances of size $d = 30$						
Heur	21.0 %	69.7 %	15.4	249.2	698.6	7200.0
Bipartite	21.6 %	68.8 %	1.3	14.6	2.0	20.0
Bipartite→Heur	20.8 %	70.0 %	13.6	716.5	15.9	776.4
Bipartite+Heur	19.4 %	72.3 %	146.7	6853.7	159.0	7200.0
GraphPool	13.7 %	81.9 %	7.4	120.1	10.6	148.6
GraphPool→Heur	13.3 %	82.6 %	21.2	446.9	27.1	508.1
GraphPool+Heur	13.2 %	82.6 %	21.6	459.9	28.8	522.7
Instances of size $d = 50$						
Heur	19.6 %	69.0 %	31.9	199.6	4404.3	21600.0
Bipartite	20.4 %	68.5 %	18.1	533.6	27.8	715.8
Bipartite→Heur	19.7 %	69.7 %	451.8	18042.9	566.1	21600.0
Bipartite+Heur	17.6 %	73.1 %	1421.1	18275.8	2086.7	21600.0
GraphPool	13.7 %	81.1 %	204.3	4875.7	291.3	6090.7
GraphPool→Heur	13.1 %	81.9 %	637.0	15768.0	813.7	17968.7
GraphPool+Heur	12.9 %	82.2 %	671.1	11768.3	1064.0	14489.8

Table 4.5: Comparing the root results for heuristic settings

thus the value of the closure of this very wide class of inequalities is not known precisely, the results suggest that the bipartite closure is a good approximation. Again it turns out that generalized MSC bipartite inequalities, which of course belong to a different class of cuts as they can't be represented as MSC inequalities for a general graph, are very effective as the root gap of GraphPool is much better than the ones of Heur, Bipartite→Heur, and Bipartite+Heur. As it was the case for Bipartite, the heuristic separation of MSC inequalities of general graphs on top of GraphPool, i.e., performing GraphPool→Heur or GraphPool+Heur, only slightly improves the root bound.

The heuristic separation of MSC inequalities has a strong effect on separation and root node processing times. After every successful cut separation, CPLEX with update the convex relaxation in order to incorporate the cut. We will refer to this as reoptimization of the relaxation. The difference between time needed for separation and overall root processing time is the portion of the time CPLEX needs to preprocess the model, solve the initial root relaxation and reoptimizations.

However, times for preprocessing and solving the initial relaxation should be identical (except random disturbances) for all approaches. Applying the heuristic to complement `Bipartite` or `GraphPool` causes a strong increase in time used for separation and reoptimization. Both, `Bipartite`→`Heur` and `Bipartite`+`Heur` need more time in the root node than `GraphPool`, while obtaining a much weaker bound.

In the approaches in the previous chapters, the separation time dominates the overall root processing time. This is not the case for `Heur` where the separation time makes up for only a negligible part of the root node processing time. One reason is that cuts for higher clique numbers are denser, i.e., contain more variables with nonzero coefficients, making the relaxations more difficult to solve. Lifting the cuts intensifies this effect, but overall has a large positive effect on the performance of the heuristic (both, in terms of faster increase of the bound and smaller overall root processing time).

The poor performance of `Heur` can also be observed on instance triangular_30_-10_0_-5__04_posDiag, which we already considered in detail earlier. The value of the bipartite closure is -3.108805 and it is obtained after 176 rounds and 17.1 seconds, where 12.7 seconds are spend for separation and 4.4 for reoptimization. `Heur` performs particularly bad on this instances. It times out after 6366 rounds with a value of -3.522827, which is substantially worst than the value of `Bipartite` as (StQP) is a minimization problem. From the 7200 seconds only 82.1 seconds are needed for separation. Figure 4.8 plots the accumulated separation and reoptimization time for `Bipartite` and `Heur` on instance triangular_30_-10_0_-5__04_posDiag for the first 400 rounds. While for `Bipartite`, the accumulated reoptimization time is smaller than the separation time and appears to grow only linear (meaning that the separation time in each round is approximately constant), the accumulated reoptimization time for `Heur` appears to grow quadratically (meaning that the relaxation becomes harder and harder to solve as the rounds proceed). Separation time for `Heur` though is very modest which shows that the overhead for clique number computations is very small.

Similar to Figs. 4.3a and 4.3b, Figs. 4.9a and 4.9b show how the dual bound changes during root node processing for triangular_30_-10_0_-5__04_posDiag. Figure 4.9a plots the dual bound against the separation round while Fig. 4.9b plots it against the time already spend at the root.

`Heur` is not nearly as effective as the other approaches. Also the hope that the additional cuts further improve the exact separators does not get fulfilled. The lines for configurations with added heuristics (dashed lines) appear to oscillate erratically around the line for the respective reference configuration. In `Bipartite`→`Heur` and `GraphPool`→`Heur` the heuristic cannot considerably improve the dual bound as the lines are almost horizontal and no effect by the heuristic is visible. After the initial steep bound improvement the configurations saturate.

Overall due to generally very limited root gap improvements and long running times of the root node, the heuristic separation of Motzkin-Straus Clique inequalities in the current form is not very effective.

4.6.5 Branch-and-cut results

After evaluating the impact of Motzkin-Straus Clique inequalities on the initial convex relaxation of (StQP), we now shift our focus toward the impact on the complete solution process. After solving the initial relaxation and, depending on the configuration, adding more inequalities to obtain a better relaxation, spatial branch-and-bound is used to solve (StQP). As spatial branching enable the separation of additional McCormick inequalities (i.e., cutting planes), the process is often also referred to as *branch-and-cut*. Most solvers are very conservative on adding cutting planes in the tree as they are might only be valid for a particular subtree and in general the trade-off between separation time and node throughput is difficult to tune. We adopt this and separate RLT and Motzkin-Straus Clique inequalities only at the root node of the tree.

Branch-and-cut results obtained with all the approaches discussed in Sections 4.6.2 and 4.6.3 are given in Table 4.6 for the case $d = 30$ and in Table 4.7 for the case $d = 50$. The tables have the same structure and report aggregated results on all the instances that can be solved to optimality by at least one of the considered approaches. For the case $d = 30$, only 3 instances cannot be solved within the time limit of 2 hours, while 10 instances with $d = 50$ are not solved by any of the methods within the time limit of 6 hours. Interestingly, all the unsolved instances are generated with positive diagonal entries and distribution $(-10, -5, 0)$. This is somewhat not surprising since for those instances the Q-space relaxation (MC-StQP) is expected to be very weak. Indeed, the negative objective coefficients drive the relaxation variables Y_{ij} towards $\min(x_i, x_j)$, which is typically much further away from the correct values of $x_i x_j$ than the opposite bound that is 0. For example, taking $x_i = x_j = \frac{1}{n}$, the correct value would be $x_i x_j = \frac{1}{n^2}$, but the linearization variables takes the value $Y_{ij} = \frac{1}{n}$. Furthermore, on these instances the diagonal terms are not linearized and thus only the weaker projected RLT inequalities can be separated instead of RLT equations.

Each of the tables gives separate results on all solved instances and on solved "hard" instances, where an instance of size $d = 30$ (resp. $d = 50$) is considered to be hard if it takes at least 30 seconds (resp. 300 seconds) to be solved with

	Solved	Time [s]		Nodes		Root Gap
		Avg.	S. Geom.	Avg.	S. Geom.	Avg.
All 119 instances solved by at least one						
CPLEX	88	2475.4	526.7	238799.2	64871.3	1006.9 %
RLT	110	585.0	26.3	17581.6	1975.2	67.2 %
Bipartite	117	362.5	20.0	7951.0	1005.4	21.7 %
GMSC	115	440.9	35.0	1420.0	468.8	11.1 %
GraphPool	117	223.0	16.9	2245.6	533.4	13.8 %
Hybrid	117	211.8	26.1	2060.4	465.3	11.4 %
HybridGP	119	189.8	24.4	1399.4	463.8	11.1 %
17 hard instances (all more than 30 seconds)						
CPLEX	2	6382.9	4864.1	442361.3	330591.4	1997.6 %
RLT	8	3989.4	1724.9	136804.2	80704.2	56.2 %
Bipartite	15	2480.0	841.3	62796.9	25174.6	20.5 %
GMSC	13	2957.3	953.2	8748.2	4481.3	13.3 %
GraphPool	15	1515.4	481.8	15902.9	8837.0	16.1 %
Hybrid	15	1384.3	363.5	14538.8	5613.0	14.6 %
HybridGP	17	1235.5	472.7	8465.9	4591.3	13.4 %

Table 4.6: Branch-and-cut results on StQPs of size 30

all compared approaches. For each of the tested methods and for each class of problems, the tables report the number of solved instances, the average computing time in seconds, the shifted geometric mean of the computing times (with a shift of 10 seconds), the average number of branch-and-bound nodes, the shifted geometric mean of the number of nodes (with a shift of 100 nodes) and the average percentage gap left at the root node. Time limits are accounted in the computations on the running time, and out-of-memory errors (that only happen for CPLEX) are accounted as time limits. Computing average and geometric mean of the number of nodes is problematic for instances that are not solved to optimality. To make a fair comparison, the calculations for number of nodes only consider those instances that all solvers but CPLEX can solve within the time limit. For CPLEX, the number of nodes processed until running out of memory or time is used and thus the reported numbers of nodes for CPLEX (which is already an order of magnitude higher as for RLT) are underestimated.

The branch-and-cut results given in Tables 4.6 and 4.7 are consistent with the ones reported in Sections 4.6.2 and 4.6.3 for the root node. On the one side, RLT inequalities appear to be fundamental for StQP, since RLT clearly

	Solved	Time [s]		Nodes		Root Gap
		Avg.	S. Geom.	Avg.	S. Geom.	Avg.
All 111 instances solved by at least one						
CPLEX	31	17834.2	13123.0	334357.7	251486.3	1455.3 %
RLT	106	1876.7	175.5	22010.8	4621.9	61.6 %
Bipartite	110	791.4	117.0	5580.8	1668.4	21.4 %
GMSC	110	841.9	203.1	2641.9	675.2	10.8 %
GraphPool	110	440.9	90.2	2647.7	687.2	14.3 %
Hybrid	110	425.5	97.9	2568.6	585.4	11.6 %
HybridGP	111	397.3	139.2	2504.5	608.2	10.9 %
17 hard instances (all more than 300 seconds)						
CPLEX	6	14497.2	7293.6	163380.6	103019.4	2105.8 %
RLT	12	10861.5	5271.1	114540.0	75263.0	63.6 %
Bipartite	16	4593.7	2389.9	28236.0	20190.5	40.1 %
GMSC	16	3960.2	2315.1	13989.6	5300.7	33.8 %
GraphPool	16	2457.2	1185.9	14369.9	5983.9	36.3 %
Hybrid	16	2354.9	1038.7	14482.7	6138.1	35.6 %
HybridGP	17	1839.3	1063.1	13279.4	3945.7	34.2 %

Table 4.7: Branch-and-cut results on StQPs of size 50

outperforms CPLEX. On the other side, MSC and GMSC bipartite inequalities are also very effective. Indeed, Bipartite outperforms RLT on all the performance indicators reported in the tables (i.e., number of solved instances, computing time and number of branch-and-bound nodes). GMSC is very effective in reducing the number of branch-and-bound nodes required and on this measure clearly even outperforms Bipartite, but running times are greater than for Bipartite indicating that separation often takes too long.

GraphPool, Hybrid, and HybridGP provide a neat improvement over Bipartite and GMSC, especially on the hard instances. This indicates that MSC bipartite inequalities are important on top of RLT inequalities and that a "clever" selection of GMSC bipartite inequalities is also important to improve over MSC bipartite inequalities. More precisely, although the number of problems solved to optimality is the same for Bipartite, GraphPool and Hybrid, separating GMSC bipartite inequalities yields a remarkable reduction in the number of nodes, which is reflected in a significant reduction in the computing times, as mentioned, especially on the hard instances. HybridGP is able to solve 3 more instances than the previously mentioned approaches and provides the best average running times,

	Solved	Time [s]		Nodes		Root Gap
		Avg.	S. Geom.	Avg.	S. Geom.	Avg.
All 119 instances solved by at least one						
Bipartite	117	362.5	20.0	4980.1	981.4	21.7 %
Bipartite→Heur	116	351.7	19.3	3797.2	857.8	20.9 %
Bipartite+Heur	117	426.9	25.6	3843.3	905.3	19.5 %
GraphPool	117	223.0	16.9	2395.7	532.9	13.8 %
GraphPool→Heur	118	214.5	20.2	2057.1	531.9	13.3 %
GraphPool+Heur	118	195.8	22.5	2002.2	560.5	13.3 %
HybridGP	119	189.8	24.4	1388.1	457.6	11.1 %
17 hard instances (all more than 30 seconds)						
Bipartite	15	2480.0	841.3	36515.6	20701.4	20.5 %
Bipartite→Heur	14	2408.5	718.2	26946.6	15142.3	19.7 %
Bipartite+Heur	15	2860.8	1018.2	24424.5	14703.8	18.4 %
GraphPool	15	1515.4	481.8	17230.3	8781.4	16.1 %
GraphPool→Heur	16	1437.7	520.9	14229.0	8044.3	15.7 %
GraphPool+Heur	16	1279.9	518.3	13323.2	7644.3	15.9 %
HybridGP	17	1235.5	472.7	8366.0	4155.4	13.4 %

Table 4.8: B&C results with heuristics size 30

but it falls behind GraphPool and Hybrid in terms of the shifted geometric mean of the running times taken over all instances. As HybridGP is compatible also in terms of shifted geometric mean on the hard instances, we conclude that the long separation times for HybridGP pay off on the hard instances, while generating too much overhead on relatively easy instances. GraphPool in contrast has the best values for shifted geometric mean over all instances and has compatible average run times so than we conclude that it is very effective in solving instances of all levels of difficulty.

Branch-and-cut results of the heuristic approached and their references are shown in Tables 4.8 and 4.9 for instances of size 30 and 50, respectively. As before, the aggregation of number of nodes only considers instances that all configurations in the table solved to optimality within the timelimit. Heur is left out as the root node results indicate it is not effective. Additionally, we compare against HybridGP as it appears to be the most effective configuration so far for hard instances and solves the largest number of instances overall.

Let us first look at the variants of Bipartite. While Bipartite→Heur improves over Bipartite in terms of time to optimality and nodes, it solves one instances

	Solved	Time [s]		Nodes		Root Gap
		Avg.	S. Geom.	Avg.	S. Geom.	Avg.
All 111 instances solved by at least one						
Bipartite	110	791.4	117.0	5611.4	1583.4	21.4 %
Bipartite→Heur	109	502.3	95.7	4125.1	1302.9	20.7 %
Bipartite+Heur	106	1482.1	176.5	3904.4	1243.9	18.4 %
GraphPool	110	440.9	90.2	2708.7	663.0	14.3 %
GraphPool→Heur	111	554.3	128.8	2739.0	692.6	13.7 %
GraphPool+Heur	110	1020.8	179.0	2855.3	773.2	13.5 %
HybridGP	111	397.3	139.2	2612.3	609.4	10.9 %
17 hard instances (all more than 300 seconds)						
Bipartite	16	4593.7	2389.9	29476.9	21210.4	40.1 %
Bipartite→Heur	16	2777.5	1408.6	20162.3	13005.8	39.7 %
Bipartite+Heur	15	5010.7	2679.9	19310.6	11288.7	37.8 %
GraphPool	16	2457.2	1185.9	15157.9	6330.5	36.3 %
GraphPool→Heur	17	2873.3	1669.7	15301.9	6880.3	35.7 %
GraphPool+Heur	17	4839.2	2888.4	15513.5	7280.9	35.5 %
HybridGP	17	1839.3	1063.1	14089.9	4281.9	34.2 %

Table 4.9: B&C results with heuristics size 50

less on both testsets. Bipartite+Heur is able to reduce the number of nodes, but show a big increase in running time, both on average and in the geometric mean and on both testsets. On the testset of size 50, Bipartite+Heur also solves 4 instances less than Bipartite. The table also shows that it is far more effective to heuristically separate GMSC bipartite inequalities as implemented in GraphPool as this configuration clearly outperforms both Bipartite→Heur and Bipartite+Heur.

The tables also show that separating cuts from general graphs in addition to GraphPool is effective, even though the picture is different for the two testsets. On instances of size 30 improve the performance in terms of average time to optimality and deteriorate it in terms of shifted geometric mean. The same holds for number of nodes. Both GraphPool+Heur and GraphPool→Heur solve on instance more than GraphPool. On the larger testset of size 50 however, applying the heuristic causes a significant increase in running time.

Overall, both ways to heuristically separate Motzkin-Straus Clique inequalities for higher clique numbers are not able to significantly improve their reference configurations. While applying the heuristic after the reference configuration is not able to separate anymore might be fruitful, applying the heuristic in every round is clearly not competitive. Clearly, non of the settings that use the heuristic is able to challenge `GraphPool` or `GraphPool+Heur` as the most effective configurations.

In order to gather more insights on the branch-and-cut results, Figs. 4.10 and 4.11 show Dolan-Moré performance profiles [DM02] on all solved instances and on solved hard instances, respectively. This time, instances of size $d = 30$ and $d = 50$ are considered together. In such plots every approach is compared to the virtual best of all approaches according to some performance measure, in our case running time. To this end, for every $x \geq 1$, the fraction of the instances where relative performance of the approach compared to the virtual best is at most x is plotted. Consequently, higher values on the y-axis (for fixed x) and smaller values on the x-axis (for fixed y) are beneficial. We omit the results obtained with default CPLEX from the plots because it is clearly dominated by all the other methods. Also `Bipartite+Heur` and `GraphPool+Heur` are not considered as they are not competitive.

The performance profiles confirm the importance of MSC and GMSC bipartite inequalities. Even from those plots one can conclude that `Bipartite` outperforms `RLT` while in turn `GraphPool` outperforms `Bipartite`. Heuristic separation in the form of `Bipartite→Heur` improves `Bipartite` but is still dominated by `GraphPool` which again indicated that GMSC bipartite inequalities boost performance much more than Motzkin-Straus Clique inequalities of graphs with higher clique numbers. The heuristic is a deterioration for `GraphPool` for about 95 % as the line of `GraphPool` dominates the one of `GraphPool→Heur` on the first part of the plot. However, `GraphPool→Heur` performs better on the hard in instances where `GraphPool` is up to a factor of 10 slower than the virtual best solver. Finally, `GraphPool` appears to be the best approach if all (solved) instances are considered (Fig. 4.10), while `Hybrid` becomes instead the best one if we restrict ourselves to hard instances only (Fig. 4.11). `HybridGP` provides the best solver for a number of hard instances and having very good average run times, however, it is clearly dominated by `GraphPool` and on a large range of factors also by `Bipartite` when all instances are considered and also not the best choice on the hard instances as it is dominated by `Hybrid`. After all, `GraphPool` appears to provide the best compromise of low overhead on instances that are easily solved by branching and strong relaxations across broad range of instances.

	Avg. Gap		Problem_30x30_0.75		Problem_50x50_0.75	
	Root	Final	Time [s]	Nodes	Time [s]	Nodes
CPLEX	4147.5 %	2220.8 %	43200.0	1370290	43200.0	276080
RLT	50.6 %	22.5 %	43.1	5130	7962.6	119634
Bipartite	11.2 %	7.9 %	12.2	1224	669.9	3239
GMSC	13.7 %	13.6 %	26.3	60	4344.6	413
GraphPool	9.2 %	8.5 %	14.9	494	513.6	440
Hybrid	8.8 %	7.6 %	31.5	63	254.0	440

Table 4.10: Results on instances from [ST08] affected by RLT or (G)MSC inequalities.

4.6.6 Computational results on the instances from [ST08]

As mentioned previously, Scozzari and Tardella [ST08] proposed a combinatorial algorithm for (StQP) and performed experiments on randomly generated instances from which a subset of 14 instances has been published. The website mentioned in the reference is no longer active, but the instances are republished on http://or.dei.unibo.it/library/msc. Since some of the instances have very large dimension (up to $d = 1000$), we impose a time limit of 12 hours.

Our callback is able to separate a cut at the root on 6 out of the 14 instances. Table 4.10 summarizes the results on these 6 instances. The first two columns shows the average gap at the root and at the timelimit of those 6 instances on which our callback separates at least one cut. Two instances (Problem_30x30_0.75 and Problem_50x50_0.75) are solved to optimality as soon as RLT inequalities are separated (i.e., in all configurations but CPLEX). Table 4.10 reports time to optimality and number of nodes for these two instances. The appendix of the online version of this thesis [Sch17] provides detailed performance figures for all instances.

The results are consistent with those of the previous sections. Although MSC and GMSC bipartite inequalities do not allow to solve more instances than RLT, they greatly reduce the root and final gap, as well as the time to optimality and the number of nodes. As before, GraphPool and Hybrid show the best compromise between time needed for separation and impact on the root gap and overall solution time. Finally, note that the 14 instances are among the hardest that the combinatorial algorithm [ST08] can solve within 1 or 2 hours of time limit, so the overall reduced gaps are quite satisfactory.

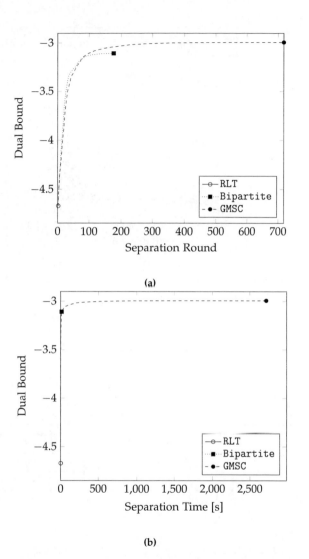

(a)

(b)

Figure 4.3: Evolution of the dual bound for instance triangular_30_-10_0_-5__04_posDiagw.r.t. separation round (a) and time (b).

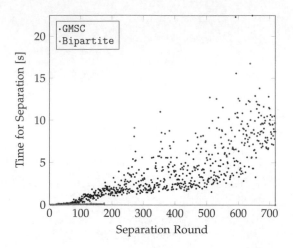

Figure 4.4: Separation time per round for instance triangular_30_-10_0_-5__04_posDiag.

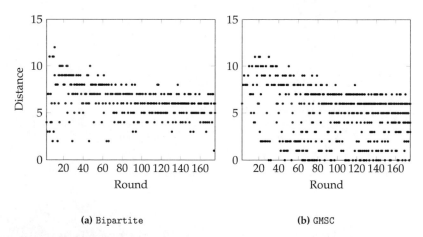

(a) Bipartite　　　　　　　　　　　　**(b)** GMSC

Figure 4.5: Distance to know graphs for instance triangular_30_-10_0_-5__04_posDiag.

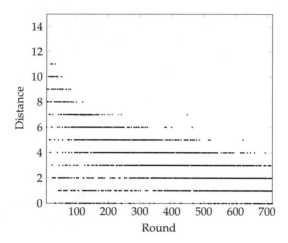

Figure 4.6: Distance to know graphs for GMSC

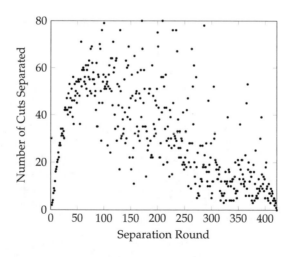

Figure 4.7: Number of cuts separated in each separation round for GraphPool on instance triangular_30_-10_0_-5__04_posDiag.

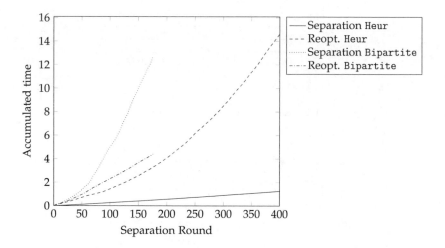

Figure 4.8: Comparison of time spent for separation and for reoptimization in the root for instance triangular_30_-10_0_-5__04_posDiag

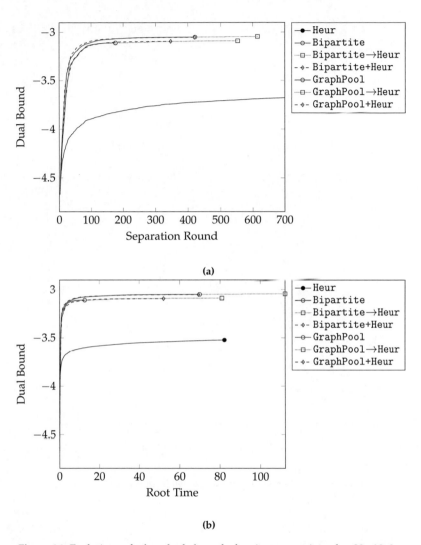

(a)

(b)

Figure 4.9: Evolution of the dual bound for instance triangular_30_-10_0_-5__04_posDiag with the exact and heuristic settings. In (a), first 700 rounds of separation shown only.

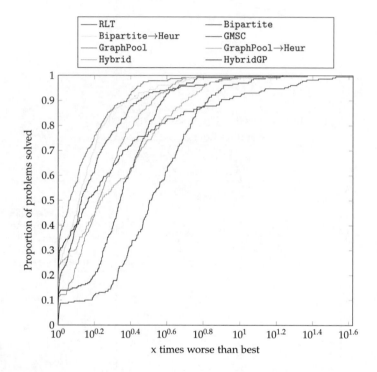

Figure 4.10: Dolan-Moré performance profile for all solved instances.

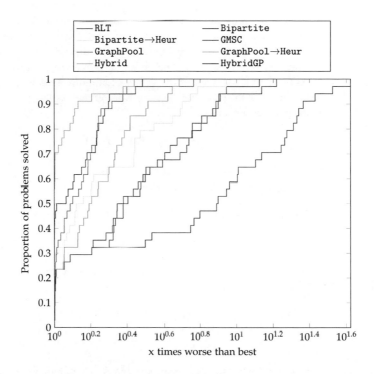

Figure 4.11: Dolan-Moré performance profile for "hard" solved instances.

4.7 Generalization

While so far we focused on the set Γ where x is in the standard simplex, we now want to generalize MSC inequalities and GMSC bipartite inequalities to more general sets. In a first step, we will show that an upper bounding inequality on the sum of the x variables suffices. In a second step, we will generalize to coefficients different than 1.

The theorem of Motzkin-Straus establishes a relation between the clique number of a graph and the optimization of a quadratic function over the standard simplex. The objective matrix is the adjacency matrix of the graph. In that QP the objective coefficients are non-negative, thus any solution with $e^T x < 1$ can be improved by a positive factor scaling, where e denotes the vector of all ones. Following this argumentation, we can relax the condition on x in the definition of the set Γ and show that MSC inequalities are valid for set

$$\Gamma_\leq = \left\{ (x, Y) \in \mathbb{R}^d \times (\mathbb{R}^d \times \mathbb{R}^d) \,\middle|\, Y = xx^T, e^T x \leq 1, x \geq 0 \right\},$$

where equation $e^T x = 1$ is relaxed to an inequality.

Lemma 4.6: *MSC inequalities are valid for all points $(x, Y) \in \Gamma_\leq$.*

Proof. Consider a graph G and a given $(x, Y) \in \Gamma_\leq$. Let

$$\beta = \frac{1}{e^T x},$$

be the scaling factor. Note that $\beta \geq 1$ is constant for fixed x. The vector $\bar{x} = \beta x$ is in the standard simplex such that

$$\langle A, Y \rangle = x^T A x \leq \beta^2 x^T A x = \bar{x}^T A \bar{x} \leq 1 - \frac{1}{\omega(G)}.$$

$\square$

The next step is to replace the simplex inequality $e^T x \leq 1$ with a more general constraint $a^T x \leq b$ and to also relax non-negativity, i.e., to approximate the set

$$\Gamma_{a,b} = \left\{ (x, Y) \in \mathbb{R}^d \times (\mathbb{R}^d \times \mathbb{R}^d) \,\middle|\, Y = xx^T, a^T x \leq b, a_i x_i \geq 0, \forall i \in V \right\}.$$

The condition $a_i x_i \geq 0$ ensures that x_i is non-negative if a_i is positive and non-positive if a_i is negative.

A similar scaling argument as above is used to generalize the MSC inequality. This time, the coefficients a and the right hand side b also determine the coefficients of the cut.

Theorem 4.7: *Let $a \in \mathbb{R}^d$, $b > 0$. Then, the following inequalities are valid for $\Gamma_{a,b}$:*

1. *Motzkin-Straus Clique inequalities*

$$\sum_{(i,j)\in E} \frac{a_i a_j}{b^2} Y_{ij} \leq 1 - \frac{1}{\omega(G)},\qquad(4.17)$$

   where $G = (V, E)$ is a simple graph.

2. *Generalized MSC bipartite inequalities*

$$\sum_{i\in M}\sum_{j\in M} \frac{a_i a_j}{b^2} Y_{ij} \leq f_\alpha\left(\sum_{i\in M} \frac{a_i}{b} x_i\right),\qquad(4.18)$$

   where $M \subset V$ and f_α is the tangent to the function $g(z) = z - z^2$ at $\alpha \in [0,1]$.

Proof. Consider $(x, Y) \in \Gamma_{a,b}$ and define $(\bar{x}, \bar{Y})$ as

$$\bar{x}_i = \frac{a_i}{b} x_i,$$
$$\bar{Y}_{ij} = \frac{a_i a_j}{b^2} Y_{ij}.$$

Node that $\bar{x} \geq 0$ and, by construction, $e^T \bar{x} \leq 1$ and $\bar{Y} = \bar{x}\bar{x}^T$ and thus $(\bar{x}, \bar{Y}) \in \Gamma_{\leq}$. Therefore, the validity of (4.17) follows directly from the previous lemma, namely

$$\sum_{(i,j)\in E} \frac{a_i a_j}{b^2} Y_{ij} = \sum_{(i,j)\in E} \bar{Y}_{ij} \leq 1 - \frac{1}{\omega(G)}.$$

To prove (4.18), it suffices to realize that the same procedure used to derive the GMSC bipartite inequalities can be repeated with $(\bar{x}, \bar{Y})$. The only difference is that we start with the inequality $e^T \bar{x} \leq 1$ instead of the equation. Multiplying this inequality by $\bar{x}_j$ gives the inequality

$$\sum_{i\in V} \bar{x}_i \bar{x}_j \leq \bar{x}_j,$$

which is valid for all $\bar{x}_j \geq 0$. All remaining operations (addition of inequalities, subtraction of terms on both sides of the inequalities) preserve the sense of the inequality such that the final result is

$$\sum_{j\in M}\sum_{i\in M} \bar{Y}_{ij} \leq f_\alpha\left(\sum_{j\in M} \bar{x}_j\right).$$

This completes the proof. $\qquad\square$

Clearly, $\Gamma_{e,1} = \Gamma_{\leq}$ and it is easy to see that the scaled MSC inequality (4.17) is really a generalization of the MSC inequality. Also, the separation problems (4.12) and (4.13) can be easily adjusted to take the coefficients a and b into account by properly scaling the solution (x^*, Y^*) to be cut off.

We end the section by noting that Theorem 4.7 establishes the applicability of Motzkin-Straus theorem to a surprisingly large family of optimization problems characterized by an indefinite quadratic objective function subject to a linear inequality. Obviously, such an inequality could also be obtained by aggregation of a large(r) system of inequalities. This makes of Motzkin-Straus Clique inequalities a rather universal tool for indefinite quadratic programming.

4.7.1 Computational experiments

In Section 4.6 MSC and GMSC bipartite inequalities have been shown to be extremely effective for StQP. However, it is not obvious if this result carries over to the generalized version of such inequalities when applied to more general problems. To start investigating this question we considered the Quadratic Knapsack Problem (QKP), which is a straightforward generalization of StQP. In QKP one asks to maximize a quadratic objective function subject to a knapsack constraint $w^T x \leq c$, where x is a vector of binary variables, w are non-negative weights and c is the non-negative capacity. We considered the continuous relaxation of QKP because CPLEX reformulates the problem to a MILP if the variables x are binary. Then, the knapsack constraint is the constraint that is used as basis to formulate RLT inequalities and (generalized) Motzkin-Straus Clique inequalities.

In our experiments we considered two sets of instances. The first set, referred as QKP1, is generated by following an approach often used in the literature, see, e.g., [CPT99; GHS80]. There, the instances are parametrized by their size d and density D, i.e., the fraction of nonzero elements in the objective function. After sampling the nonzero elements, the objective coefficients $Q_{ij} = Q_{ji}$ are sampled uniformly from $[1, 100]$. The weight w_i are sampled uniformly from $[1, 50]$ and the capacity c from $[50, \sum_{i=1}^{d} w_i]$. As for StQP, we use only fully dense objective matrices, i.e., $D = 1.0$, and we generated 150 instances with size $d = 30$. The second set of instances, referred as QKP2 in the following, has the same structure (again 150 instances), but the objective matrices are sampled as described in Section 4.6.1. Instances from both test sets are available on http://or.dei.unibo.it/library/msc.

Computations were done for the four configurations CPLEX, RLT, Bipartite and Hybrid that are defined as in Section 4.6.2. In this case, GMSC turns out to be too expensive computationally. In these experiments, another cutting plane

	CPLEX	RLT	Bipartite	Hybrid
BQP disabled				
Average root gap [%]				
Gap left	5.65	0.29	0.11	0.10
Closed wrt CPLEX root	–	80.26	85.21	85.41
Closed wrt RLT	–	–	39.62	40.84
Solved/Timeout at the root				
Solved	6	25	39	39
Timeout	0	0	0	12
Affected				
RLT/MSC/GMSC	0/0/0	144/0/0	144/101/0	144/101/47
BQP enabled				
Average root gap [%]				
Gap left	5.65	0.02	0.00	0.00
Closed wrt CPLEX root	–	94.80	95.74	95.74
Closed wrt RLT	–	–	24.24	24.24
Solved/Timeout at the root				
Solved	6	110	144	143
Timeout	0	0	0	2
Affected				
RLT/MSC/GMSC	0/0/0	144/0/0	144/103/0	144/101/5

Table 4.11: Root node results on Quadratic Knapsack instances QKP1.

technique already applied by CPLEX, namely BQP cuts [BGL16; IBMa], has a substantial impact. We present results both with BQP cuts disabled, to get a fair comparison between closures, and with BQP cuts enabled, to evaluate the effect of combined cutting planes.

Tables 4.11 and 4.12 show aggregated results at the root node on instances QKP1 and QKP2, respectively. The tables are split into two parts: with BQP cuts disabled and enabled. For each of the considered settings, the tables report the average root gaps, the number of instances solved, and the number of instances where the root node processing was not finished with the timelimit and, finally, the number of instances that are affected by each class of inequalities. An instance is considered to be affected by a given class if at least one cut from that class is separated.

	CPLEX	RLT	Bipartite	Hybrid
BQP disabled				
Average root gap [%]				
Gap left	124.86	118.44	116.40	115.74
Closed wrt CPLEX root	–	22.69	28.14	29.12
Closed wrt RLT	–	–	16.08	17.78
Solved/Timeout at the root				
Solved	24	32	43	46
Timeout	0	0	0	0
Affected				
RLT/MSC/GMSC	0/0/0	94/0/0	94/72/0	94/72/73
BQP enabled				
Average root gap [%]				
Gap left	48.77	45.29	44.63	44.54
Closed wrt CPLEX root	–	27.66	37.05	39.76
Closed wrt RLT	–	–	22.39	26.48
Solved/Timeout at the root				
Solved	29	40	55	56
Timeout	0	0	0	21
Affected				
RLT/MSC/GMSC	0/0/0	94/0/0	94/80/0	94/72/69

Table 4.12: Root node results on Quadratic Knapsack instances QKP2.

As seen for StQP, applying RLT inequalities is very beneficial, both in terms of number of instances solved at the root and of root gap reduction. The effect is more pronounced on the instances of type QKP1 (Table 4.11) where 144 out of 150 instances are affected by RLT, 19 additional instances are solved w.r.t. CPLEX when BQP cuts are disabled and 104 with BQP cuts enabled. With BQP cuts disabled, the average gap left is reduced from 5.65 to 0.29 %, Bipartite can then close an additional 39.62 % gap w.r.t. RLT leaving only 0.11 % on average and solving in total 39 instances at the root. Surprisingly, separating GMSC bipartite inequalities on top of Bipartite has almost no effect. Even though almost one third of the instances is affected, number of instances solved and gap stay (almost) the same. When the various types of cuts are combined with BQP cuts, the overall comparison is similar but the gap closed by all techniques is even more important.

Remarkably, Bipartite and Hybrid can solve almost all instances (with only 6 and 7 unsolved instances at the root, respectively).

Instances of type QKP2 (Table 4.12) have much larger average root gap, but show similar phenomena. With and without BQP cuts, RLT and Bipartite contribute to the solution of several additional instances at the root. With BQP cuts disabled, the effect of Bipartite on the average root gap is very small, but 16 % gap is closed with respect to RLT. Even if Hybrid allows to solve 3 more instances w.r.t. Bipartite, GMSC bipartite inequalities appear to be less effective w.r.t. what we observed for StQP where the impact on the root gap is much more marked. Again the combination with BQP cuts seems beneficial.

The appendix of the online version of this thesis [Sch17] provides detailed performance figures for all instances.

4.8 Conclusion

We studied cutting planes for standard quadratic programs and quadratic knapsack. Those cutting planes exploit the relation between those problems and the maximum clique problem. By analyzing the relationship between the new cutting planes and the RLT inequalities, we have shown that, interestingly, (*i*) MSC bipartite inequalities are not comparable with first level RLT inequalities, and (*ii*) the derivation of GMSC bipartite inequalities generalizes both MSC bipartite inequalities and first level RLT. Our computational experiments show that both MSC bipartite inequalities and GMSC bipartite inequalities allow to get a significantly stronger bound than first level RLT alone.

Some possible extensions of our approach would be to exploit generalized versions of the Motzkin-Straus Theorem [Gib+97].

5 Strong Relaxations for the Pooling Problem

The pooling problem is a classic non-convex, nonlinear problem introduced by Haverly in 1978 [Hav78]. In a nutshell, the task is to route flow through a network. The raw materials at the inputs exhibit known qualities of with respect to certain attributes and at the outputs bound on the qualities have to be respected. The point is that at each node of the network, the arriving material gets blended and therefore, the attribute qualities have to be tracked across the network. We formally introduce the problem in Section 5.1. The pooling problem has important application in the petrochemical industry, e.g., [QG95], where crude oil of different quality and composition from several oil fields or seeder vessels has to be stored in tanks and then blended into end-products with certain characteristics for the downstream network. A similar use-case are wastewater treatment networks, e.g., [GG98], and mining, e.g., [Bol+15]. [CKM16] developed an approach to identify pooling substructures in general MINLP, so that techniques that have been developed specifically for the pooling problem can be automatically transferred to other domains where the pooling structure is hidden in a bigger problem formulation. Motivated by these applications, in this chapter we use the structure of the pooling problem to derive strong convex inequalities that strengthen the relaxation of the state-of-the-art formulation for this problem.

Several formulations for this problem have been proposed in the literature. All of them include bilinear terms to model repeated blending of the material, i.e., the situation where some part of the incoming flow is not coming directly from in input but has been blended before. The traditional formulations along with their corresponding McCormick relaxations are presented in Section 5.2.

At this point it comes to no surprise that tight linear or convex relaxations for non-convex problems are of great importance for the global solution process and that improvements to the McCormick relaxation can be achieved by taking the problem structure into account. Using the structure of the pooling problem is also the key to the set of relaxations we present in Section 5.3.1. First, the well-known pq-formulation for the pooling problem is extended to a higher space by introducing variables that are linear aggregations of original problem variables. Then, the original variables are projected out and only five variables, their bound constraints, four linear constraints and one bilinear constraint remain. This small set forms a relaxation of the pooling problem, but still retains some of the non-convex essence of the problem. Due to the bilinear constraint, the relaxation is still non-convex, but its small size allows to study its convex hull. To this end, we first describe all extreme points of the convex hull in Section 5.3.2. The extreme points have been helpful to identify the valid inequalities that are presented in Section 5.3.3. It turns out that even though the structure of the set is fairly simple, its study and the description of the convex hull in involved. We prove that the proposed inequalities indeed describe the convex hull of the non-convex relaxation for certain values of the input parameters in Section 5.3.4. Finally, Section 5.4 presents computational experiments that show the impact of the valid inequalities on the relaxation value and within a global solution approach. Section 5.5 concludes this chapter.

This chapter is joint work with Claudia D'Ambrosio, Jeff Linderoth, and Jim Luedtke. Parts of it were conducted during a research stay of the author at École Polytechnique in Paris within the MINO project. A publication that will contain large parts of this chapter, in particular Sections 5.3 and 5.4, is in preparation.

5.1 Introduction

The pooling problem consists of finding an optimal flow of different raw materials through a network considering that the materials mix at several points in the network. The raw materials have different attributes whose concentration is constrained at the output. Pooling networks consist of three layers of nodes: Inputs, pools, and outputs. Figure 5.1 visualizes a pooling network. Flow is only allowed from inputs to pools, pools to outputs and directly from inputs to outputs as visualized by the solid arcs in Fig. 5.1. Material with potentially different attribute concentrations gets blended at the pools and at the outputs. The necessity to track the attribute concentrations across the network is the main difference between a regular network flow problem and accounts for the difficulty of the pooling problem as the fact that the blended material from the pools is again blended at the outputs is responsible for the nonconvexity of the problem. A problem of this form is called *standard pooling problem*. The pooling problem has important applications in the petrochemical industry and also in water networks.

Even if the problem is known since decades, only in 2013 it was proved to be strongly NP-hard [AH13a]. Further complexity results on special cases of the

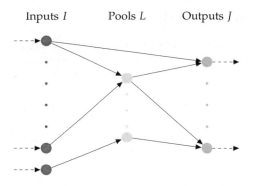

Inputs I Pools L Outputs J

Figure 5.1: The pooling problem seen as an optimization problem on a graph.

pooling problem where obtained more recently, see [Hau15; BKR15].

In this thesis, we rely on the McCormick relaxation (see Chapter 3) as means of convexification and a spatial branching approach (see Section 2.2.3) to ensure global optimality; an approach that has be used in numerous publications on the topic, e.g., [AH13a; Aud+04; BEG94; QG95; TS02; RC09; Gup12; MGF10; Rui+12; FHJ92].

Different types of approaches to tackle with the pooling problem can be found in the literature, from the methods based on recursive or successive linear programming problems, see [Hav78; BL85], on decompositions, see [FA90; BEG94], or on Lagrangian relaxation [ATS99; AE08]. [GMF09] describes a piecewise-linear relaxation. Note also that several variants of the standard pooling problem have been considered, see, for example, the *generalized pooling problem* [AH13b; Aud+04; RC09; BKR16], where flow is also allowed between pools, or pooling problems with additional constraints [Rui+12; MGF10]. Standard and generalized pooling problems typically are continuous problem. In the presence of design decisions that are modeled by integer variables the model becomes an MINLP and is called *extended pooling problem*, see, for example, [Vis01; MF06; MF10; DLL11]. A multi-period extension of the pooling problem is found in [Lot+16]. [Bol+15] consider a variation with multiple periods and the additional twist that target concentrations are soft constraints and derivations from the target concentrations are penalized in the objective function. [Gup12; DG15] employ discretization techniques to approximate the problem and as a heuristic. Several authors describe heuristics for the pooling problem, e.g., [Aud+04; AH14]. For comprehensive overviews the reader is referred to [TS02; RC09; Gup12; Gup+15].

5.2 Standard formulations

The fundamental difference between different formulations is the tracking of concentration of the attributes in the pool. The p-formulation is the most natural model and uses dedicated variables for the concentration for each attribute in each pool. The q-formulation in contrast does not explicitly compute the concentration in the pool but in the spirit of a multi-commodity for problem tracks the flow from the different inputs through the network. Knowing how the end product decomposes into the different input streams allows to compute the concentrations at the outputs. Both models can not avoid bilinear terms as they have to ensure that the concentrations are correctly computed at the output nodes and in particular taking into account the contributions from the pools. The pq-formulation, finally, is a strengthening of the q-formulation.

5.2.1 Notation and assumptions

Before we start introducing the formulations, we introduce the problem parameters. Table 5.1 summarizes the notation used to describe the parameters of a pooling problem and its formulations.

The network structure is modeled as a directed graph $G = (V, A)$. Each node belongs to one of the three sets: inputs I, pools L, and outputs J. Inputs, pools and outputs partition the nodes such that

$$V = I \cup L \cup J.$$

In this part of the thesis, we will use the indices i for inputs, ℓ for pools and j for outputs. Generic nodes will be denoted by u and v.

Flow is only allowed from inputs to pools, from pools to outputs and as direct connection between inputs and outputs. Accordingly, the arc set A is thus composed of three subsets

$$A = A_{IL} \cup A_{LJ} \cup A_{IJ}$$

with

$$A_{IL} := \{(i, \ell) \in A \mid i \in I, \ell \in L\}$$
$$A_{LJ} := \{(\ell, j) \in A \mid \ell \in L, j \in J\}$$
$$A_{IJ} := \{(i, j) \in A \mid i \in I, j \in J\}.$$

We chose to disallow connections between pools for notational convenience and refer to [Gup12] for the modifications needed in the presence of connections between pools.

Item	Domain	Meaning
Sets		
I		Inputs
L		Pools
J		Outputs
I_u	$\subset I$	Inputs with direct connection to node u
L_u	$\subset L$	Pools with direct connection to node u
J_u	$\subset J$	Outputs with direct connection to node u
K		Attribute
V		All nodes in the network
A		All arcs in the network
A_{IL}	$\subset A$	Arcs from inputs to pools
A_{LJ}	$\subset A$	Arcs from pools to outputs
A_{IJ}	$\subset A$	Arcs from inputs to outputs
Indices		
i	$\in I$	An input
ℓ	$\in L$	A pool
j	$\in J$	An output
k	$\in K$	An attribute
Parameters		
C_u	$\in [0, \infty)$	Capacity of node u
C_{uv}	$\in [0, \infty)$	Capacity of arc (u, v)
λ_{ik}	$\in [0, 1]$	Concentration of attribute k and input i
$\underline{\mu}_{jk}$	$\in [0, 1]$	Lower bound on the concentration of attribute k at output j
$\overline{\mu}_{jk}$	$\in [0, 1]$	Upper bound on the concentration of attribute k at output j
$\underline{\gamma}_{ijk}$	$\in [0, 1]$	Excess of concentration of attribute k in material coming from input w.r.t. lower bound at output j
$\overline{\gamma}_{ijk}$	$\in [0, 1]$	Excess of concentration of attribute k in material coming from input w.r.t. upper bound at output j
Variables		
x_{uv}	$\in [0, C_{uv}]$	Flow from node u to node v
p_{uk}	$\in [0, 1]$	Concentration of attribute k at node u
$q_{i\ell}$	$\in [0, 1]$	Fraction of the flow through pool ℓ coming from input i
$w_{i\ell j}$	$\in \mathbb{R}_{\geq 0}$	Flow from input i through pool ℓ to output j

Table 5.1: Notation for the pooling problem and its formulations

To formulate the models, it is useful to know the set of nodes that have a direct connection to a specific node. We therefore define the sets

$$I_u := \{i \in I \mid (i, u) \in A\}$$
$$L_u := \{\ell \in L \mid (\ell, u) \in A \text{ or } (u, \ell) \in A\}$$
$$J_u := \{j \in J \mid (u, j) \in A\}$$

While arcs can only leave input nodes and enter output nodes, both can happen for pools.

Nodes and arcs have flow capacities. The capacity of node u is denoted with $C_u \in [0, \infty)$. We set $C_u = \infty$ is not bound is present. For inputs and outputs the interpretation of the capacity is easy to understand as maximum amount that can be inserted and extracted from the network. This might be a technical limitation or due to limitations in upstream or downstream networks. The capacity C_ℓ of the pool ℓ can be interpreted as the size of the tank where blending happens and limits the amount of flow through the pool. The capacity $C_{u,v} \in [0, \infty)$ of an arc (u, v) is the maximum amount of flow that is allowed over this arc.

Until now we only introduced the graph structure used to model pooling problems. A pooling problem however is more than a flow problem on a graph. The twist comes from the restrictions on the concentrations of the attributes. We consider a set K of attributes that are tracked. We use the index k to refer to a specific attribute in K. Depending on the application and the attribute it can be a desirable or an undesirable property. The concentration of the different attributes is known at the inputs. The quantity $\lambda_{ik} \in [0, 1]$ is the fraction of attribute $k \in K$ in the material that is fed in at $i \in I$. Certain minimum and maximum concentrations at the outputs have to be respected. The values $\underline{\mu}_{jk}, \overline{\mu}_{jk} \in [0, 1]$ with $\underline{\mu}_{jk} \leq \overline{\mu}_{jk}$ represent lower and upper bounds on the concentration of attribute $k \in K$ at output $j \in J$.

We assume linear blending. In this blending model, in each node the contribution of each incoming stream of material to the concentration of an attribute is proportional to the contribution of the flow to the total incoming flow. In other words, the concentration of some attribute $k \in K$ is computed as the weighted average of the incoming concentrations with the weights being the fraction of the total flow incoming from a specific incoming arc.

5.2.2 The flow model

The underlying problem structure of the pooling problem is a network flow problem. The flow along the arcs in the network is the crucial information of the solution; all other values can be derived from them. Since flow variables and the associated constraints are present in every model formulation, we present it first.

A flow variable x_{ij} models the flow along each arc $(i, j) \in A$. The flow is constraint by upper and lower bounds

$$0 \leq x_{uv} \leq C_{uv} \qquad \text{for all } (u, v) \in A. \qquad (5.1)$$

Nonnegativity expresses that flow is only allowed from node u to node v, but not in the reverse direction. Furthermore, the flow might have a finite upper bound C_{uv}.

While flow enters the network at inputs and leaves at output, no flow enters or leaves at pools. Standard flow conservation constraints that couple incoming and outgoing flow ensure this condition:

$$\sum_{i \in I_\ell} x_{i\ell} = \sum_{j \in J_\ell} x_{\ell j} \qquad \text{for all } \ell \in L \qquad (5.2)$$

The capacity C_ℓ of the pool ℓ limits the flow through the pool, so is an upper bound on any of the two sums. The capacity constraints then read

$$\sum_{i \in I_\ell} x_{i\ell} \leq C_\ell \qquad \text{for all } \ell \in L. \qquad (5.3)$$

Capacities also have to be respected at inputs and outputs. There, the capacity is an upper bound on the flow leaving input to the connected pools and outputs and and entering an output from the connected inputs and pools, respectively:

$$\sum_{\ell \in L_i} x_{i\ell} + \sum_{j \in J_i} x_{ij} \leq C_i \qquad \text{for all } i \in I \qquad (5.4)$$

$$\sum_{\ell \in L_j} x_{\ell j} + \sum_{i \in I_j} x_{ij} \leq C_j \qquad \text{for all } j \in J \qquad (5.5)$$

The left hand side of (5.4) computes the sum leaving input i to pools and outputs. At the same time, this sum is the amount of upstream material that is feed into the network at i. Equivalently, the left hand side of (5.5) is the amount of flow leaving the pooling network to the downstream at output j.

As mentioned before, flow variables and constraints are present in all model formulations. Differences in the formulations come from the handling of the attribute tracking. With only the flow variables, it is impossible to formulate bounds on the concentrations at the output nodes. Sections 5.2.3 and 5.2.4 present two different ways to model tracking of the concentrations and formulate the respective constraints.

5.2.3 The p-formulation

The p-formulation was introduced by Haverly when he introduced the pooling problem in [Hav78] and is arguable the most intuitive one. It relies on explicitly computing the concentration of each attribute in the pools and in the outputs. While the model is straight-forward and explicit, its drawback is the poor relaxation it provides.

In order to explicitly compute the attribute concentrations at the pools and outputs, concentration variables $p_{\ell k}, p_{jk} \in [0,1]$ are introduced for each pool $\ell \in L$, output $j \in J$ and attribute $k \in K$.

Due to linear blending, the concentration of some attribute $k \in K$ at a pool $\ell \in L$ is the weighted average of the incoming concentrations

$$p_{\ell k} = \sum_{i \in I_\ell} \lambda_{ik} \frac{x_{i\ell}}{\sum_{\bar{i} \in I_\ell} x_{\bar{i}\ell}} \tag{5.6}$$

Of course, (5.6) is only valid for non-zero total inflow $\sum_{\bar{i}:(\bar{i},\ell) \in A_{IL}} x_{\bar{i}\ell}$. In case the total inflow is zero, $p_{\ell k}$ is allowed to take an arbitrary value.

Equation (5.6) looks overly complicated since indeed it can be reformulated into

$$p_{\ell k} \sum_{i \in I_\ell} x_{i\ell} = \sum_{i \in I_\ell} \lambda_{ik} x_{i\ell}. \tag{5.7}$$

For the outputs, the situation is slightly more involved. First, flow is coming in from inputs and pools. Second, the concentration of the material coming from the pools is itself a variable and yields bilinear constraints as we will see.

Following the same route as for pools, we first represent the concentration of an attribute $k \in K$ at an output $j \in J$ as the weighted sum

$$p_{jk} = \frac{\sum_{i \in I_j} \lambda_{ik} x_{ij} + \sum_{\ell \in L_j} p_{\ell k} x_{\ell j}}{\sum_{i \in I_j} x_{ij} + \sum_{\ell \in L_j} x_{\ell j}} \tag{5.8}$$

Again, p_{jk} can be chosen arbitrary if the total inflow in the denominator is zero. Multiplying with the denominator gives the final constraint:

$$p_{jk} \left(\sum_{i \in I_j} x_{ij} + \sum_{\ell \in L_j} x_{\ell j} \right) = \sum_{i \in I_j} \lambda_{ik} x_{ij} + \sum_{\ell \in L_j} p_{\ell k} x_{\ell j} \tag{5.9}$$

Equations (5.7) and (5.9) are often referred to as *spec tracking constraints*, since they model correct tracking of the attribute in presence of mixing in the pools and at the outputs.

At this point, all building blocks of the p-formulation have been stated and we restate the entire model. We use a generic minimization objective function f, since a large variety of objectives, such as maximization of profits from selling output goods, minimization of raw material purchasing costs, and penalties for flow running over specific arcs, are possible. In our computational experiments, we assign costs to flow over each arc, so that

$$f(x, p) = \sum_{(u,v) \in A} c_{uv} x_{uv}$$

where the coefficients c_{uv} are part of the input data. The p-formulation of the pooling problem then is:

$$\min \quad f(x, p) \tag{5.10a}$$

$$\text{s.t.} \quad \sum_{i \in I_\ell} x_{i\ell} = \sum_{j \in J_\ell} x_{\ell j} \leq C_\ell \qquad \text{for all } \ell \in L \tag{5.10b}$$

$$\sum_{\ell \in L_i} x_{i\ell} + \sum_{j \in J_i} x_{ij} \leq C_i \qquad \text{for all } i \in I \tag{5.10c}$$

$$\sum_{\ell \in L_j} x_{\ell j} + \sum_{i \in I_j} x_{ij} \leq C_j \qquad \text{for all } j \in J \tag{5.10d}$$

$$p_{\ell k} \sum_{i \in I_\ell} x_{i\ell} = \sum_{i \in I_\ell} \lambda_{ik} x_{i\ell} \qquad \text{for all } \ell \in L, \ k \in K \tag{5.10e}$$

$$p_{jk} \left(\sum_{i \in I_j} x_{ij} + \sum_{\ell \in L_j} x_{\ell j} \right) = \sum_{i \in I_j} \lambda_{ik} x_{ij} + \sum_{\ell \in L_j} p_{\ell k} x_{\ell j} \quad \text{for all } j \in J, \ k \in K \tag{5.10f}$$

$$\underline{\mu}_{jk} \leq p_{jk} \leq \overline{\mu}_{jk} \qquad \text{for all } j \in J, \ k \in K \tag{5.10g}$$

$$0 \leq x_{uv} \leq C_{uv} \qquad \text{for all } (u, v) \in A \tag{5.10h}$$

Bilinear terms appear in (5.10e) and (5.10f) and are always products of one p and one x variable. As these variables are bounded by (5.10g) and (5.10h) the McCormick relaxation (see Section 3.2.1) is easily constructed and we refer to it as the *p-relaxation*.

One property that is common to all formulations of the pooling problem is that the termwise McCormick relaxation already provides the sharpest convexification of all multilinear terms involved. In the p-relaxation, consider the term

$$p_{\ell k} \sum_{i \in I_\ell} x_{i\ell}$$

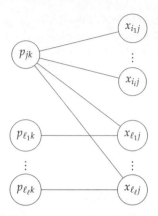

Figure 5.2: The support graph of the expression (5.11)

from (5.10e). The support graph is a bipartite graph and the coefficients are all positive. Thus by Corollary 3.2, the McCormick relaxation provides the convex envelope and as there are no negative coefficients the McCormick relaxation also provides the concave envelope. In (5.10f), the expression is

$$p_{jk} \left(\sum_{i \in I_j} x_{ij} + \sum_{\ell \in L_j} x_{\ell j} \right) - \sum_{\ell \in L_j} p_{\ell k} x_{\ell j}. \tag{5.11}$$

Also here the support graph is bipartite as products are always between p and x variables, but the coefficients are neither all positive nor all negative such that Corollary 3.2 does not apply. We thus argue that the support graph has no cycle and apply Corollary 3.3. Figure 5.2 shows the support graph of (5.11) for a some $j \in J$ and $k \in K$. The nodes corresponding to the variables x_{ij} for $i \in I_j$ and $p_{\ell k}$ for $\ell \in L_j$ have degree one as the respective variables appear only once in the expression. The remainder of the graph consists of the single variable p_{jk} which is connected to all $x_{\ell j}$ for $\ell \in L_j$. This star graph clearly has no cycle.

Overall, the McCormick relaxation of the bilinear expressions in this formulation cannot be tightened without using more problem structure.

5.2.4 The q-formulation

In contrast to the p-formulation, the q-formulation avoids the explicit computation of pool and output concentrations of the attributes. Instead, it uses variables to express the proportion of the total flow though a pool that is coming from a particular input to track the origin of flow. The restriction on the concentration of the attributes can then be formulated without the need to have explicit variables even for the output concentrations. The q-formulation was introduced in [BEG94].

The first step towards the q-formulation is the introduction of proportion variables $q_{i\ell} \in [0,1]$ that express the proportion of the total flow though pool ℓ originating from input i. In order to properly express a proportion, the variables for each pool have belong to a standard simplex. Namely, apart from being nonnegative, the variables belonging to one pool $\ell \in L$ have to sum up to one:

$$\sum_{i \in I_\ell} q_{i\ell} = 1 \tag{5.12}$$

Proportion variables make it easy to track the paths the flow from a particular input takes through the network. Let $w_{i\ell j}$ be the flow along the path from input i though pool ℓ into output j. Consequently, we call the variables $w_{i\ell j}$ *path variables*. Clearly, $w_{i\ell j}$ is just the product of the flow from ℓ to j with the proportion of the flow in the pool coming from i:

$$w_{i\ell j} = q_{i\ell} x_{\ell j} \tag{5.13}$$

The flow $x_{i\ell}$ from an input i to a pool ℓ can then also be expressed in terms for path variables by summing the path variables that go from i through ℓ to all different outputs

$$x_{i\ell} = \sum_{j \in J_\ell} w_{i\ell j} \tag{5.14}$$

Equations (5.12)–(5.14) ensure a consistent flow model. Indeed it is a well known fact that these constraints are already sufficient to ensure the flow conservation constraints. We prove it here for completeness.

Theorem 5.1: *Equations* (5.12)–(5.14) *imply the flow conservation constraints* (5.2) *for all pools.*

Proof. Consider a fixed pool $\ell \in L$. We start by summing up the constraints (5.2) for all inputs with connection to the pool and then use the definition of the path variables.

$$\sum_{i \in I_\ell} x_{i\ell} = \sum_{i \in I_\ell} \sum_{j \in J_\ell} w_{i\ell j}$$

$$= \sum_{i \in I_\ell} \sum_{j \in J_\ell} q_{i\ell} x_{\ell j}$$

$$= \sum_{i \in I_\ell} \left(q_{i\ell} \sum_{j \in J_\ell} x_{\ell j} \right)$$

$$= \left(\sum_{i \in I_\ell} q_{i\ell} \right) \left(\sum_{j \in J_\ell} x_{\ell j} \right)$$

$$= \sum_{j \in J_\ell} x_{\ell j}$$

The last equation is due to (5.12) and completes the proof. $\qquad\square$

The proof uses the nonlinear definition (5.13) of the path variables. Adding the flow conservation constraints to the model thus might strengthen the McCormick relaxation as this identity does not necessarily hold. In the next section, however, we introduce the pq-formulation as a strengthening of the q-formulation and will show that for the pq-formulation the flow conservation constraints are also redundant for the McCormick relaxation.

The next step is to compute the concentration of a attribute $k \in K$ at a given output $j \in J$. Flow and path variables allow direct access to the amount of flow that is going from any input to output j. Again, the concentration at the output is the weighted average of the incoming concentrations and the requirement at j for attribute k can be formulated as

$$\underline{\mu}_{jk} \leq \frac{\sum_{i \in I_j} \lambda_{ik} x_{ij} + \sum_{\ell \in L_j} \sum_{i \in I_\ell} \lambda_{ik} w_{i\ell j}}{\sum_{i \in I_j} x_{ij} + \sum_{\ell \in L_j} \sum_{i \in I_\ell} w_{i\ell j}} \leq \overline{\mu}_{jk} \tag{5.15}$$

which of course can be linearized by multiplying the inequality with the denominator.

For notational convenience, we choose to make one more reformulation and formulate the concentration requirement in terms of excess over the two bounds. To this end we introduce the excess of the concentration in the material from an input i with respect to the bounds at output j for attribute k as

$$\underline{\gamma}_{ijk} = \lambda_{ik} - \underline{\mu}_{jk}$$

$$\overline{\gamma}_{ijk} = \lambda_{ik} - \overline{\mu}_{jk}.$$

$\overline{\gamma}_{ijk}$ and $\underline{\gamma}_{ijk}$ measure the excess of the concentration at the input with respect to upper and lower quality bounds at the output, respectively. Upper and lower bounding inequalities now need to be formulated separately to replace constraint (5.15).

$$\sum_{i \in I_j} \underline{\gamma}_{ijk} x_{ij} + \sum_{\ell \in L_j} \sum_{i \in I_\ell} \underline{\gamma}_{ijk} w_{i\ell j} \geq 0 \tag{5.16}$$

$$\sum_{i \in I_j} \overline{\gamma}_{ijk} x_{ij} + \sum_{\ell \in L_j} \sum_{i \in I_\ell} \overline{\gamma}_{ijk} w_{i\ell j} \leq 0 \tag{5.17}$$

These are all building blocks of the q-formulation of the pooling problem which we state next:

$$\min \quad f(x, q) \tag{5.18a}$$

$$\text{s.t.} \quad \sum_{j \in J_\ell} x_{\ell j} \leq C_\ell \qquad \text{for all } \ell \in L \tag{5.18b}$$

$$\sum_{\ell \in L_i} x_{i\ell} + \sum_{j \in J_i} x_{ij} \leq C_i \qquad \text{for all } i \in I \tag{5.18c}$$

$$\sum_{\ell \in L_j} x_{\ell j} + \sum_{i \in I_j} x_{ij} \leq C_j \qquad \text{for all } j \in J \tag{5.18d}$$

$$\sum_{i \in I_\ell} q_{i\ell} = 1 \qquad \text{for all } \ell \in L \tag{5.18e}$$

$$w_{i\ell j} = q_{i\ell} x_{\ell j} \qquad \text{for all } i \in I_\ell, \, \ell \in L_j, \, j \in J \tag{5.18f}$$

$$x_{i\ell} = \sum_{j \in J_\ell} w_{i\ell j} \qquad \text{for all } i \in I_\ell, \, \ell \in L \tag{5.18g}$$

$$\sum_{i \in I_j} \underline{\gamma}_{ijk} x_{ij} + \sum_{\ell \in L_j} \sum_{i \in I_\ell} \underline{\gamma}_{ijk} w_{i\ell j} \geq 0 \qquad \text{for all } j \in J, \, k \in K \tag{5.18h}$$

$$\sum_{i \in I_j} \overline{\gamma}_{ijk} x_{ij} + \sum_{\ell \in L_j} \sum_{i \in I_\ell} \overline{\gamma}_{ijk} w_{i\ell j} \leq 0 \qquad \text{for all } j \in J, \, k \in K \tag{5.18i}$$

$$q_{i\ell} \geq 0 \qquad \text{for all } i \in I_\ell, \, \ell \in L \tag{5.18j}$$

$$0 \leq x_{uv} \leq C_{uv} \qquad \text{for all } (u, v) \in A \tag{5.18k}$$

Notice that the nonlinearity has been isolated into (5.18f) and all other constraints are linear. Of course, (5.18f) can be used to replace every occurrence of $w_{i\ell j}$ and the question arises whether the resulting bilinear expressions can be better convexified as by the termwise McCormick relaxation. However, as for the p-formulation, this is not the case. The three equations where $w_{i\ell j}$ occurs are (5.18g)–(5.18i). Clearly, the graph corresponding to the bilinear expression in (5.18g) is bipartite and all coefficients are positive such that Theorem 3.1 and Corollary 3.2 ensures that termwise McCormick gives the best convexification of that expression. For (5.18h) and (5.18i) the situation is more complicated as the sign of the concentration excess $\overline{\gamma}_{ijk}$ is not known. The corresponding

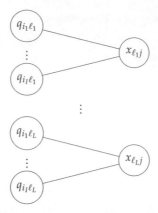

Figure 5.3: The support graph of the expression (5.19)

bilinear expression reads

$$\sum_{\ell \in L_j} \sum_{i \in I_\ell} \overline{\gamma}_{ijk} q_{i\ell} x_{\ell j}. \tag{5.19}$$

and its support graph is visualized in (5.19). Focusing on the inner sum, we see that the output j and the pool ℓ are fixed and that the variable $x_{\ell j}$ is connected to all proportion variables $q_{i\ell}$ for this pool. The result is a star graph which clearly has no cycles. Thus every inner loop contributes a star graph to the support graph of the expression. The outer loop creates a collection of star graphs, but they are not connected as the outer loop is over the pools ℓ and the pool is an index for all the variables. Overall, a graph where every connected component is a star clearly has no cycles and by Corollary 3.3 also for this formulation, the McCormick relaxation provides the best possible convexification for the bilinear expressions without considering more problem structure.

5.2.5 The *pq*-formulation

The *pq*-formulation was introduced in [TS02] and is considered one of the strongest formulations for the pooling problem. The *pq*-formulation arises from the *q*-formulation by application of RLT (see Section 3.3). Recall that in RLT, equations or inequalities from the model are multiplied with variables or constraints to get a valid constraint involving higher order terms. These higher order terms are then linearized by reformulating them with appropriate linearization variables. Typically, this is done if the resulting higher order terms are already part of the model and thus the linearization variables already exit. In this light, the two constraints (5.12) and (5.3) are suitable for being used in the RLT procedure.

Equation (5.12) involves only $q_{i\ell}$ variables for one pool $\ell \in L$ and all inputs $i \in I_\ell$ that connect to the pool. Multiplied by any flow variable $x_{\ell j}$ for the pool and any of the connected outputs $j \in J_\ell$, the result involves terms $q_{i\ell}x_{\ell j}$ which are linearized by $w_{i\ell j}$. The additional constraints for this combination of variables and constraints are thus

$$\sum_{i \in I_\ell} w_{i\ell j} = x_{\ell j} \qquad \text{for all } j \in J_\ell, \ell \in L. \qquad (5.20)$$

Since the original constraint was an equation, the additional one is also an equation.

Equation (5.3) in turn only involves flow variables $x_{\ell j}$ for one pool $\ell \in L$ and all connected outputs $j \in J_\ell$ and is thus suitable to be multiplied by any proportion variable $q_{i\ell}$ for this pool and any connected input $i \in I_\ell$ that is connected to the pool ℓ. The additional constraints for this combination of variables and constraints are thus

$$\sum_{j \in J_\ell} w_{i\ell j} \leq C_\ell q_{i\ell} \qquad \text{for all } i \in I_\ell, \ell \in L \qquad (5.21)$$

In this case, the result is an inequality since the original constraint was an inequality which was multiplied by a nonnegative variable.

While the flow conservation constraints are implied for feasible solutions of the p-formulation, they are not implied for its relaxation. With the additional constraints on the path variables, they are now also implied for the relaxation. They are easily derived using just linear equations:

$$\sum_{i \in I_\ell} x_{i\ell} = \sum_{i \in I_\ell} \sum_{j \in J_\ell} w_{i\ell j}$$
$$= \sum_{j \in J_\ell} \sum_{i \in I_\ell} w_{i\ell j}$$
$$= \sum_{j \in J_\ell} x_{\ell j}$$

Clearly, both constraints (5.20) and (5.21) are redundant. However, they significantly strengthen the McCormick relaxation making the pq-formulation one of the strongest formulations known and the starting point for constructing stronger relaxations in the next section.

5.3 New convex relaxations for the pooling problem

As the relaxations obtained by the formulations described in the previous section can not be easily improved by better approximating the bilinear terms, more problem structure has to be used to obtain stronger relaxations for the pooling problem. The idea is to start from the pq-formulation as the formulation with the strongest relaxation and again create a relaxation thereof. All inequalities that are valid for the relaxation are then also valid for the pq-formulation. We therefore seek a set that is rich enough in the structure of the pooling problem, but small enough for an in-depth study. To this end, we focus on a very small subset of the features of an instances of the pooling problem, precisely 1 attribute, 1 output, and 1 pool, and ignore all constraints without direct relation to these features. The size of the relaxation is further reduced by aggregating variables in a linear fashion and project the problem on these aggregated variables. The final relaxation will consist of only 5 variables, 1 equation involving a bilinear term, 4 linear constraints and bounds on some of the variables, yet it still captures some of the essential structure of the pooling problem, i.e., the nonconvexity introduced by equations involving bilinear terms. At the same time its very small size allows a study of its structure. More precisely, we study of its convex hull as we seek a strong convex relaxation of the pooling problem by adding convex inequalities.

The roadmap for Section 5.3 is to first develop the relaxation in Section 5.3.1. In Section 5.3.2 we describe all extreme points of the convex hull of the relaxation. Section 5.3.3 contains the new valid inequalities for the pooling problem along with proofs of their validity. Finally, in Section 5.3.4 we show that these inequalities suffice to describe the convex hull of the relaxation developed in Section 5.3.1 for certain parameters of the pooling problem.

5.3.1 A 5 variable relaxation

Intuitively, the constraints (5.18h) and (5.18i) on the concentrations of attributes at the outputs are of central importance for the pooling problem. Essentially, ensuring feasible output concentrations is the reason for tracking the concentrations across the network and without these two constraints, the problem would reduce to a rather simple flow problem.

The first step towards the desired relaxation is to focus on one particular inequality (5.18h) or (5.18i) and extract a set that still obeys this inequality, but relaxes all other inequalities of this class. The simplifies the problem considerably. The inequalities (5.18h) and (5.18i) have the same structure (We can always multiply (5.18h) -1 to also get a $\leq$-inequality), so the relaxation can be constructed by keeping any of them. So, first fix an attribute $\hat{k} \in K$ and output $\hat{j} \in J$ and remove all constraints (5.18h) and (5.18i) on the concentrations except (5.18i) for $\hat{k}$ and $\hat{j}$ from the model. Furthermore, we remove all capacity constraints (5.18b)–(5.18d) except the one for output $\hat{j}$. As the path variables $w_{i\ell j}$ are only needed to track the flow from input to output, they now irrelevant for all outputs $j \neq \hat{j}$ and we relax all constraints where path variables $w_{i\ell j}$ for $j \neq \hat{j}$ occur, in particular (5.18g) and (5.21). Overall, we only use the constraints (5.18e) and (5.18j) and the constraints (5.18d), (5.18f), and (5.18i) for the fixed output $\hat{j}$ and attribute $\hat{k}$. The first relaxation is then:

$$\sum_{\ell \in L_{\hat{j}}} x_{\ell\hat{j}} + \sum_{i \in I_{\hat{j}}} x_{i\hat{j}} \leq C_{\hat{j}} \tag{5.22a}$$

$$\sum_{i \in I_{\hat{j}}} \overline{\gamma}_{ij\hat{k}} x_{i\hat{j}} + \sum_{\ell \in L_{\hat{j}}} \sum_{i \in I_{\ell}} \overline{\gamma}_{ij\hat{k}} w_{i\ell\hat{j}} \leq 0 \tag{5.22b}$$

$$\sum_{i \in I_{\ell}} q_{i\ell} = 1 \qquad \text{for all } \ell \in L_{\hat{j}} \tag{5.22c}$$

$$\sum_{i \in I_{\ell}} w_{i\ell\hat{j}} = x_{\ell\hat{j}} \qquad \text{for all } \ell \in L_{\hat{j}} \tag{5.22d}$$

$$w_{i\ell\hat{j}} = q_{i\ell} x_{\ell\hat{j}} \qquad \text{for all } i \in I_{\ell},\ \ell \in L_{\hat{j}} \tag{5.22e}$$

$$q_{i\ell} \geq 0 \qquad \text{for all } i \in I_{\ell},\ \ell \in L_{\hat{j}} \tag{5.22f}$$

$$x \geq 0 \tag{5.22g}$$

Nonnegativity of x (5.22g) is assumed to hold for all indices that appear in the model.

While already much smaller, the model still has too many non-convex constraints (5.22e). In the next step, we fix one pool $\hat{\ell} \in L$ and aggregate all flows that don't flow through that pool $\hat{\ell}$. We call the aggregated value *bypass flow*. This includes flow through other pools and also through any direct connection between an input and $\hat{j}$.

More precisely, we introduce a new variable $z_{\hat{\ell}\hat{j}}$ which measures the bypass flow and define it by

$$z_{\hat{\ell}\hat{j}} = \sum_{i \in I_{\hat{j}}} x_{i\hat{j}} + \sum_{\substack{\ell \in L_{\hat{j}} \\ \ell \neq \hat{\ell}}} x_{\ell\hat{j}} \tag{5.23}$$

The concentration of the material going through the bypass is not tracked exactly, but is bounded with conservative bounds based on the entries that can send flow though the bypass. Note that all terms from the left hand side of (5.22a) are used except for $x_{\hat{\ell}\hat{j}}$. As all flows are nonnegative, we also ask nonnegativity from $z_{\hat{\ell}\hat{j}}$.

The variable for the weighted concentration excess in the bypass is denoted by $y_{\hat{\ell}}$ and is defined by

$$y_{\hat{\ell}\hat{j}k} = \sum_{i\in I_{\hat{j}}} \overline{\gamma}_{ijk} x_{ij} + \sum_{\substack{\ell\in L_{\hat{j}} \\ \ell\neq\hat{\ell}}} \sum_{i\in I_\ell} \overline{\gamma}_{ijk} w_{i\ell\hat{j}}. \tag{5.24}$$

Similar to above, the variable $y_{\hat{\ell}\hat{j}k}$ contains almost all terms from the left hand side of (5.22b), namely all except for

$$\sum_{i\in I_{\hat{\ell}}} \overline{\gamma}_{ijk} w_{i\hat{\ell}\hat{j}}.$$

The concentration of the flow through the bypass is bounded by the highest and lowest concentrations among all reachable inputs, so that the following bounds for $y_{\hat{\ell}\hat{j}k}$ are valid:

$$\underline{\beta}_{\hat{\ell}\hat{j}k} z_{\hat{\ell}\hat{j}} \leq y_{\hat{\ell}\hat{j}k} \leq \overline{\beta}_{\hat{\ell}\hat{j}k} z_{\hat{\ell}\hat{j}} \tag{5.25}$$

$$\underline{\beta}_{\hat{\ell}\hat{j}k} = \min(\min_{i\in I_{\hat{j}}} \overline{\gamma}_{ijk}, \min_{\substack{\ell\in L_{\hat{j}} \\ \ell\neq\hat{\ell}}} \min_{i\in I_\ell} \overline{\gamma}_{ijk}) \tag{5.26}$$

$$\overline{\beta}_{\hat{\ell}\hat{j}k} = \max(\max_{i\in I_{\hat{j}}} \overline{\gamma}_{ijk}, \max_{\substack{\ell\in L_{\hat{j}} \\ \ell\neq\hat{\ell}}} \max_{i\in I_\ell} \overline{\gamma}_{ijk}) \tag{5.27}$$

With these new variables, (5.22) is further relaxed. First, the new variables are used to reformulate (5.22a) and (5.22b). Furthermore, as only the fixed pool $\hat{\ell}$ is modeled explicitly, the constraints (5.22c)–(5.22e) all omitted for all $\ell \neq \hat{\ell}$. Variables referring to any pool $\ell \neq \hat{\ell}$ don't appear in the reduced model anymore and are also projected out. The relaxed model then reads:

$$z_{\hat{\ell}\hat{j}} + x_{\hat{\ell}\hat{j}} \leq C_{\hat{j}} \tag{5.28a}$$

$$y_{\hat{\ell}\hat{j}k} + \sum_{i\in I_{\hat{\ell}}} \overline{\gamma}_{ijk} w_{i\hat{\ell}\hat{j}} \leq 0 \tag{5.28b}$$

$$\underline{\beta}_{\hat{\ell}\hat{j}k} z_{\hat{\ell}\hat{j}} \leq y_{\hat{\ell}\hat{j}k} \leq \overline{\beta}_{\hat{\ell}\hat{j}k} z_{\hat{\ell}\hat{j}} \tag{5.28c}$$

$$\sum_{i\in I_{\hat{\ell}}} q_{i\hat{\ell}} = 1 \tag{5.28d}$$

$$\sum_{i\in I_{\hat{\ell}}} w_{i\hat{\ell}\hat{j}} = x_{\hat{\ell}\hat{j}} \tag{5.28e}$$

$$w_{i\hat{\ell}\hat{j}} = q_{i\hat{\ell}} x_{\hat{\ell}\hat{j}} \qquad \text{for all } i \in I_{\hat{\ell}} \tag{5.28f}$$

$$q_{i\hat{\ell}} \geq 0 \qquad \text{for all } i \in I_{\hat{\ell}} \tag{5.28g}$$

$$x_{\hat{\ell}\hat{j}} \geq 0 \tag{5.28h}$$

$$z_{\hat{\ell}\hat{j}} \geq 0 \tag{5.28i}$$

The relaxation (5.28) still has $|I_{\hat{\ell}}|$ non-convex equations (5.28f). We therefore go one step further and also aggregate all inputs that are connected to $\hat{\ell}$. By (5.28e), $x_{\hat{\ell}\hat{j}}$ describes the amount of flow that goes from all inputs though the pool $\hat{\ell}$ to $\hat{j}$. In addition, we introduce two aggregated variables:

$$u_{\hat{\ell}\hat{j}\hat{k}} = \sum_{i \in I_{\hat{\ell}}} \overline{\gamma}_{ij\hat{k}} w_{i\hat{\ell}\hat{j}} \tag{5.29}$$

$$t_{\hat{\ell}\hat{k}} = \sum_{i \in I_{\hat{\ell}}} q_{i\hat{\ell}} \overline{\gamma}_{ij\hat{k}} \tag{5.30}$$

The variable $u_{\hat{\ell}\hat{j}\hat{k}}$ describes the contribution to the weighted excess in the quality constraint (5.28b) that is coming from the pool $\hat{\ell}$. The variable $t_{\hat{\ell}\hat{k}}$ uses the proportion variables to model the excess in concentration w.r.t. $\hat{j}$ of the material in the pool.

$x_{\hat{\ell}\hat{j}}$, $u_{\hat{\ell}\hat{j}\hat{k}}$, and $t_{\hat{\ell}\hat{k}}$ are connected by a non-convex relation:

$$
\begin{aligned}
u_{\hat{\ell}\hat{j}\hat{k}} &= \sum_{i \in I_{\hat{\ell}}} \overline{\gamma}_{ij\hat{k}} w_{i\hat{\ell}\hat{j}} \\
&= \sum_{i \in I_{\hat{\ell}}} \overline{\gamma}_{ij\hat{k}} q_{i\hat{\ell}} x_{\hat{\ell}\hat{j}} \\
&= x_{\hat{\ell}\hat{j}} \sum_{i \in I_{\hat{\ell}}} \overline{\gamma}_{ij\hat{k}} q_{i\hat{\ell}} \\
&= x_{\hat{\ell}\hat{j}} t_{\hat{\ell}\hat{k}}
\end{aligned}
\tag{5.31}
$$

Equation (5.31) comes to no surprise. $u_{\hat{\ell}\hat{j}\hat{k}}$ measures the contribution of flow going through the pool in the concentration restriction for $\hat{k}$ and is computed as the weighted excess with the weight being the flow values from the inputs to the fixed output $\hat{j}$. Equation (5.31) says this should be the same as the excess in the pool (i.e., $t_{\hat{\ell}\hat{k}}$) multiplies with the flow from the pool to the exit (i.e., $x_{\hat{\ell}\hat{j}}$).

For the McCormick relaxation of (5.31), bounds of both variables are needed. Bounds for $x_{\hat{\ell}\hat{j}}$ are easily obtained from (5.28a) and (5.28h):

$$0 \leq x_{\hat{\ell}\hat{j}} \leq C_{\hat{j}}$$

The best easily computable bounds for $t_{\hat{\ell}\hat{k}}$ are the minimum and maximum input concentrations from inputs feeding pool $\hat{\ell}$. A lower bound is easily derived by

$$
\begin{aligned}
t_{\hat{\ell}\hat{k}} &= \sum_{i \in I_{\hat{\ell}}} q_{i\hat{\ell}} \overline{\gamma}_{ij\hat{k}} \\
&\leq \sum_{i \in I_{\hat{\ell}}} q_{i\hat{\ell}} \min_{i' \in I_{\hat{\ell}}} \overline{\gamma}_{i'j\hat{k}} \\
&= \min_{i \in I_{\hat{\ell}}} \overline{\gamma}_{ij\hat{k}}
\end{aligned}
$$

The last equation is because the minimum is independent of index i and the proportion variables sum up to 1. An upper bound is derived the same way such that bounds on $t_{\hat{\ell}\hat{k}}$ are:

$$\min_{i \in I_{\hat{\ell}}} \overline{\gamma}_{ijk} \leq t_{\hat{\ell}\hat{k}} \leq \max_{i \in I_{\hat{\ell}}} \overline{\gamma}_{ijk} \tag{5.32}$$

In the final step to construct a relaxation that is amendable to in-depth study, we use the new variable $u_{\hat{\ell}\hat{j}\hat{k}}$ to reformulate (5.28b) and add the valid relationship (5.31). Finally, explicitly computing the proportion and path variables is not needed, so we don't consider (5.28d)–(5.28g) anymore and project out any unused variables.

Before stating the proposed relaxation, we want to stress that the relaxation can be constructed for every choice of $\hat{i} \in I$, $\hat{\ell} \in L$, and $\hat{k} \in K$. The remaining variables are $t_{\hat{\ell}\hat{k}}$, $u_{\hat{\ell}\hat{j}\hat{k}}$, $x_{\hat{\ell}\hat{j}}$, $y_{\hat{\ell}\hat{j}\hat{k}}$, and $z_{\hat{\ell}\hat{j}}$. To simplify the notation, we will from now on drop these indices and denote the lower and upper bounds on $t_{\hat{\ell}\hat{k}}$ by $\underline{\gamma}$ and $\overline{\gamma}$, respectively. The relaxation then reads:

$$u - xt = 0 \tag{5.33a}$$
$$y + u \leq 0 \tag{5.33b}$$
$$z + x \leq C \tag{5.33c}$$
$$y \leq \overline{\beta}z \tag{5.33d}$$
$$y \geq \underline{\beta}z \tag{5.33e}$$
$$z \geq 0$$
$$x \in [0, C]$$
$$t \in [\underline{\gamma}, \overline{\gamma}]$$

We denote the set of points that fulfill the above inequalities and equations with T (again omitting that indices $\hat{i}$, $\hat{\ell}$, and $\hat{k}$ that T depends on).

By properly scaling the variables but t, we can assume w.l.o.g. that $C = 1$ and will do so from now on. Interestingly, even though starting with the pq-formulation, all variables that characterize this formulation, namely proportion and path variables, have been projected out. Model (5.33) instead has an explicit variable for the concentration of the fixed attribute in the pool. The model is therefore much more in the spirit of the p-formulation.

Note that all definitions of the additional variables are linear in the original variables. Thus, any linear or convex inequality valid for T can be "lifted" to a linear or convex inequality in the original variables of the pooling problem.

Due to the nonlinear equation $u = xt$, T is still a non-convex set. Using the bounds $0 \leq x \leq 1$ and $\underline{\gamma} \leq t \leq \overline{\gamma}$, the constraint $u = xt$ can be relaxed by the

McCormick inequalities:

$$u - \underline{\gamma}x \geq 0 \tag{5.34}$$

$$\overline{\gamma}x - u \geq 0 \tag{5.35}$$

$$u - \underline{\gamma}x \leq t - \underline{\gamma} \tag{5.36}$$

$$\overline{\gamma}x - u \leq \overline{\gamma} - t \tag{5.37}$$

Equations (5.34)–(5.37) provide the best possible convex relaxation of the feasible points of $u = xt$ given that x and t are in the bounds mentioned above. However, not all such triples of (u, x, t) that fulfill (5.34)–(5.37) meet the requirements expressed by the other constraints in the model. Taking the entire constraint set into account to construct a convex relaxation of T is the objective of this work. Clearly, the convex hull conv(T) would give the best convex relaxation of T and our aim is to provide a complete description of conv(T) which in general is not a polyhedron.

Note that (5.34)–(5.37) imply the bounds $0 \leq x \leq 1$ and $\underline{\gamma} \leq t \leq \overline{\gamma}$. Also the bound constraint $z \geq 0$ is implied by (5.33d) and (5.33e). Thus, we define the standard relaxation of the set T by

$$R^0 := \{(x, u, y, z, t) \,|\, (5.33b)–(5.33e) \text{ and } (5.34)–(5.37)\}.$$

Every convex set is described completely by its extreme points and rays. T is bounded such that no extreme rays are present. In the following section, we characterize the extreme point of conv(T).

5.3.2 Extreme points

In this section, a complete list of the extreme points of conv(T) will be provided. Recall that a point $p \in T$ is extreme if it can not be represented as convex combination of two district points from the set, i.e., if there are not two other points $p_1, p_2 \in T$ with $p_1 \neq p_2$ and a $\lambda \in (0,1)$ with $p = \lambda p_1 + (1 - \lambda)p_2$. The set of extreme points of conv(T) is denoted by ext(conv(T)). Note that ext(conv(T)) $\subset T$.

The following lemma states that extreme points either transport no material at all ($z + x = 0$) or that the capacity of the output is fully used ($z + x = 1$).

Lemma 5.2: *If $p = (x, t, z, y, u) \in$ ext(conv(T)), then $z + x \in \{0, 1\}$.*

Proof. Suppose there was a $p = (x, t, z, y, u) \in$ ext(conv(T)) with $0 < z + x < 1$. Then $p \in T$ and there exists an $\epsilon > 0$ such that scaling all variable except t with $(1 + \epsilon)$ and scaling them with $(1 - \epsilon)$ yields two feasible points $p_+ \in T$ and $p_- \in T$. Since $p = 0.5p_+ + 0.5p_-$, p is not extreme which is a contradiction. $\square$

When there is no flow though the network, the concentration in the pool can take arbitrary values in the bounds. The result are two extreme points for this case.

Theorem 5.3: *The points $p = (x,t,z,y,u) \in \text{ext}(\text{conv}(T))$ with $z + x = 0$ are $(0,\underline{\gamma},0,0,0)$ and $(0,\overline{\gamma},0,0,0)$.*

Proof. Since extreme points are in T, the condition $z + x = 0$ implies $z = x = u = y = 0$. The only variable that is not fixed is t which can vary freely in its bound interval $[\underline{\gamma},\overline{\gamma}]$. Hence the two extreme points. $\qquad\square$

By Lemma 5.2, all remaining extreme points fulfill $z + x = 1$ and Lemma 5.4 reveals the structure of them.

Lemma 5.4: *If $p = (x,t,z,y,u) \in \text{ext}(\text{conv}(T))$ and $z + x = 1$, then either*

- $x = 1$,

- $z = 1$, *or*

- $y + u = 0$.

Proof. Let $p = (x,t,z,y,u) \in \text{ext}(\text{conv}(T))$ with $x < 1$, $z < 1$, and $y + u < 0$. Let K such that $y = Kz$. Define

$$p_+ = (x(1+\epsilon),t,1-x(1+\epsilon),K(1-x(1+\epsilon)),tx(1+\epsilon))$$
$$p_- = (x(1-\epsilon),t,1-x(1-\epsilon),K(1-x(1-\epsilon)),tx(1-\epsilon))$$

Since $p \in T$, both p_+ and p_-, are in T and $p = 0.5p_+ + 0.5p_-$. Therefore, p is not extreme and not all three inequalities can be strict simultaneously. $\qquad\square$

The following two theorems characterize the extreme points with $z = 1$ and with $x = 1$.

Theorem 5.5: *If $\underline{\beta} > 0$, then $\text{ext}(\text{conv}(T))$ has no extreme points with $z = 1$. If $\underline{\beta} \leq 0$, then the points $p = (x,t,z,y,u) \in \text{ext}(\text{conv}(T))$ with $z = 1$ are*

- $(0,\underline{\gamma},1,\underline{\beta},0)$ *and* $(0,\overline{\gamma},1,\underline{\beta},0)$ *and*

- $(0,\underline{\gamma},1,\min(\overline{\beta},0),0)$ *and* $(0,\overline{\gamma},1,\min(\overline{\beta},0),0)$.

Proof. Since $z = 1$, we have $x = u = 0$, $y \in [\underline{\beta},\overline{\beta}] \cap [-\infty,0]$, and $t \in [\underline{\gamma},\overline{\gamma}]$. This set becomes infeasible if $[\underline{\beta},\overline{\beta}] \cap [-\infty,0] = \emptyset$, i.e., if $\underline{\beta} > 0$. If $\underline{\beta} \leq 0$, the extreme points are the vertices of the hyperrectangle $\{0\} \times [\underline{\gamma},\overline{\gamma}] \times \{1\} \times [\underline{\beta},\min(\overline{\beta},0)] \times \{0\}$. $\qquad\square$

Theorem 5.6: *If* $\underline{\gamma} > 0$, *then* $\text{ext}(\text{conv}(T))$ *has no extreme points with* $x = 1$.
If $\underline{\gamma} \leq 0$, *then the points* $p = (x, t, z, y, u) \in \text{ext}(\text{conv}(T))$ *with* $x = 1$ *are*

- $(1, \underline{\gamma}, 0, 0, \underline{\gamma})$ *and*

- $(1, \min(\overline{\gamma}, 0), 0, 0, \min(\overline{\gamma}, 0))$.

Proof. If $x = 1$, then $z = y = 0$, $u = t$. Since $t = u \leq 0$ and $t \in [\underline{\gamma}, \overline{\gamma}]$ the system is infeasible if $\underline{\gamma} > 0$ and there are not extreme points with $x = 1$. Otherwise, i.e., if $\underline{\gamma} \leq 0$, the extreme points with $x = 1$ are completely characterized by $t = \underline{\gamma}$ and $t = \min(\overline{\gamma}, 0)$. $\qquad\square$

The remaining extreme points fulfill $z + x = 1$ and $y + u = 0$. Using these two equations and propagating them through the defining inequalities of T, it is easy to see that the the remaining extreme points have only two liberties and satisfy the following system which we denote by $\tilde{T}$:

$$-xt \leq \overline{\beta}(1 - x) \qquad (5.38)$$
$$-xt \geq \underline{\beta}(1 - x) \qquad (5.39)$$
$$x \in [0, 1]$$
$$t \in [\underline{\gamma}, \overline{\gamma}].$$

The remaining variables can be computed by substitution and other choices as x and t as variables describing the two liberties are possible. All extreme points of $\text{conv}(\tilde{T})$ are extreme points of $\text{conv}(T)$.

To attack $\tilde{T}$, we fix the variable t to $t = \alpha$ and define the set

$$T_\alpha := \left\{ (x, t, z, y, u) \in T \,\middle|\, z + x = 1,\, y + u = 0,\, t = \alpha \in [\underline{\gamma}, \overline{\gamma}] \right\}.$$

It is clear that $\text{conv}(T) = \text{conv}(\left\{ T_\alpha \,\middle|\, \alpha \in [\underline{\gamma}, \overline{\gamma}] \right\})$. Now, it is not clear anymore that all extreme points of T_α are extreme points of $\text{conv}(\tilde{T})$; some might actually lie in the interior of $\text{conv}(\tilde{T})$. We know, however, that the extreme point of T with $z + x = 1$ and $y + u = 0$ are in $\cup_{\alpha \in [\underline{\gamma}, \overline{\gamma}]} \text{ext}(T_\alpha)$.

Theorem 5.7: *The extreme points of* T_α, $\alpha \notin \{\underline{\beta}, \overline{\beta}\}$, *are completely characterized by*

- $x = \max(0, \frac{\underline{\beta}}{\underline{\beta} - \alpha})$ *and* $x = \min(1, \frac{\overline{\beta}}{\overline{\beta} - \alpha})$ *if* $\alpha < \underline{\beta} < \overline{\beta}$

- $x = 0$ *and* $x = \min(1, \frac{\underline{\beta}}{\underline{\beta} - \alpha}, \frac{\overline{\beta}}{\overline{\beta} - \alpha})$ *if* $\underline{\beta} < \alpha < \overline{\beta}$

- $x = \max(0, \frac{\overline{\beta}}{\overline{\beta} - \alpha})$ *and* $x = \min(1, \frac{\underline{\beta}}{\underline{\beta} - \alpha})$ *if* $\underline{\beta} < \overline{\beta} < \alpha$

Proof. The set T_α has only one degree of freedom and every $p \in T_\alpha$ is of the form

$$p = (x, \alpha, 1 - x, -x\alpha, x\alpha)$$

for some x.

Working with (5.38), we get

$$x(\overline{\beta} - \alpha) \leq \overline{\beta} \qquad \Longleftrightarrow \qquad x \begin{cases} \leq \frac{\overline{\beta}}{\overline{\beta} - \alpha} & \text{if } \overline{\beta} - \alpha > 0 \\ \geq \frac{\overline{\beta}}{\overline{\beta} - \alpha} & \text{if } \overline{\beta} - \alpha < 0 \end{cases}$$

For $\overline{\beta} - \alpha = 0$, the system is infeasible if $\overline{\beta} < 0$ and feasible with $x \in [0,1]$ if $\overline{\beta} \geq 0$. (In any case, $\alpha = \overline{\beta}$ does not yield new extreme points for conv(T), since the extreme points in this case are on the line between two previously known extreme points.)

Similar for (5.39)

$$x(\underline{\beta} - \alpha) \geq \underline{\beta} \qquad \Longleftrightarrow \qquad x \begin{cases} \geq \frac{\underline{\beta}}{\underline{\beta} - \alpha} & \text{if } \underline{\beta} - \alpha > 0 \\ \leq \frac{\underline{\beta}}{\underline{\beta} - \alpha} & \text{if } \underline{\beta} - \alpha < 0 \end{cases}$$

Here, for $\underline{\beta} - \alpha = 0$, the system is infeasible if $\underline{\beta} > 0$ and feasible with $x \in [0,1]$ if $\underline{\beta} \leq 0$. (Also in this case we don't get new extreme points for conv(T).)

With the condition $\alpha \notin \{\underline{\beta}, \overline{\beta}\}$, we get three cases:

case $\alpha < \underline{\beta} < \overline{\beta}$:	case $\underline{\beta} < \alpha < \overline{\beta}$:	case $\underline{\beta} < \overline{\beta} < \alpha$:
$x \leq \dfrac{\overline{\beta}}{\overline{\beta} - \alpha}$	$x \leq \dfrac{\overline{\beta}}{\overline{\beta} - \alpha}$	$x \geq \dfrac{\overline{\beta}}{\overline{\beta} - \alpha}$
$x \geq \dfrac{\underline{\beta}}{\underline{\beta} - \alpha}$	$x \leq \dfrac{\underline{\beta}}{\underline{\beta} - \alpha}$	$x \leq \dfrac{\underline{\beta}}{\underline{\beta} - \alpha}$

which completes the proof. $\qquad\qquad\qquad\qquad\qquad\qquad\qquad\qquad\qquad\qquad\square$

Theorems 5.3, 5.5 and 5.6 provide a discrete set of extreme points of conv(T) and Theorem 5.7 provides a continuous set of points that might or might not be extreme for conv(T).

5.3.3 Valid inequalities

We next derive four additional classes of valid inequalities for conv(T), two of them linear, and two of them convex nonlinear. As we will see, depending on the signs of $\underline{\gamma}$ and $\overline{\gamma}$, some of these inequalities are redundant.

Theorem 5.8: *If $\underline{\beta} < 0$, then the following inequality is valid for T:*

$$(u - \underline{\beta}x)(u - \underline{\gamma}x) \leq -\underline{\beta}x(t - \underline{\gamma}) \tag{5.40}$$

Proof. Aggregating the inequalities (5.33b) (with weight 1), (5.33c) (with weight $-\underline{\beta}$), and (5.33e) (with weight 1) yields the following valid inequality for R^0:

$$u - \underline{\beta}x \leq -\underline{\beta}.$$

Multiplying this inequality by $x(t - \underline{\gamma}) \geq 0$ on both sides yields the nonlinear inequality

$$(u - \underline{\beta}x)x(t - \underline{\gamma}) \leq (-\underline{\beta})x(t - \underline{\gamma})$$

which is also valid for R^0. Substituting $u = xt$ into the left-hand-side of this yields (5.40). $\qquad\square$

We observe that if $\overline{\gamma} < 0$, then (5.40) is redundant. Indeed, $\overline{\gamma} < 0$ implies $t < 0$ and therefore $u < 0$, which in turn implies $u - \underline{\beta}x < -\underline{\beta}x$. In the other hand $x \leq 1$ and $t - \underline{\gamma} > 0$ imply that $t - \underline{\gamma} \geq xt - \underline{\gamma}x = u - \underline{\gamma}x$. Furthermore, $0 = u - xt \leq u - \underline{\gamma}x$ and $-\underline{\beta}x \geq 0$. Combining all that we get that (5.40) is always strict if $\overline{\gamma} < 0$:

$$(u - \underline{\beta}x)(u - \underline{\gamma}x) < -\underline{\beta}x(u - \underline{\gamma}x)$$
$$\leq -\underline{\beta}x(t - \underline{\gamma})$$

We next show that (5.40) is second-order cone representable and thus convex. We can rewrite (5.40) as:

$$(u - \underline{\beta}x)(u - \underline{\gamma}x) \leq -\underline{\beta}x(t - \underline{\gamma})$$
$$\Leftrightarrow (u - \underline{\gamma}x)^2 + (\underline{\gamma} - \underline{\beta})x(u - \underline{\gamma}x) \leq -\underline{\beta}x(t - \underline{\gamma})$$
$$\Leftrightarrow (u - \underline{\gamma}x)^2 \leq x\left[(-\underline{\beta})(t - \underline{\gamma}) + (\underline{\beta} - \underline{\gamma})(u - \underline{\gamma}x)\right]$$

This is a rotated second-order cone $2x_1x_2 \geq x_3^2$ in dimension 3 with the auxiliary variables

$$x_1 = \frac{1}{2}s$$
$$x_2 = (-\underline{\beta})(t - \underline{\gamma}) + (\underline{\beta} - \underline{\gamma})(u - \underline{\gamma}x)$$
$$x_3 = u - \underline{\gamma}x.$$

It is left to show is that $x_1, x_2 \geq 0$, which is immediately clear for x_1. The following lemma shows the nonnegativity of x_2 and therefore the second-order cone representability of (5.40).

Lemma 5.9: *If $\underline{\beta} < 0$, the following inequality is valid for T:*

$$(-\underline{\beta})(t - \underline{\gamma}) + (\underline{\beta} - \underline{\gamma})(u - \underline{\gamma}x) \geq 0 \tag{5.41}$$

Proof. First, as $\underline{\beta} < 0$ then by $y + u \leq 0$, $-\underline{\beta}(x + z) \leq -\underline{\beta}$, $\underline{\beta}z - y \leq 0$, and $\underline{\gamma}x - u \leq 0$, we have:

$$(\underline{\gamma} - \underline{\beta})x \leq -\underline{\beta}$$

and therefore, using $t - \underline{\gamma} \geq 0$,

$$(\underline{\gamma} - \underline{\beta})(t - \underline{\gamma})x \leq (-\underline{\beta})(t - \underline{\gamma}). \tag{5.42}$$

But then, using $u = tx$, yields

$$(\underline{\gamma} - \underline{\beta})(t - \underline{\gamma})x = (\underline{\gamma} - \underline{\beta})(tx - \underline{\gamma}x) = (\underline{\gamma} - \underline{\beta})(u - \underline{\gamma}x).$$

Substituting into (5.42) yields:

$$(\underline{\gamma} - \underline{\beta})(u - \underline{\gamma}x) \leq (-\underline{\beta})(t - \underline{\gamma})$$

and rearranging yields the result. $\qquad\square$

The second inequality is only conditionally valid.

Theorem 5.10: *If $\overline{\beta} > 0$ and $\underline{\gamma} < 0$, then the following inequality is valid for T when $y > 0$:*

$$(\overline{\gamma} - \underline{\gamma})y + \overline{\beta}(\overline{\gamma}x - u) + \frac{\underline{\gamma}y(u - \underline{\gamma}x)}{y + u - \underline{\gamma}x} \leq \overline{\beta}(\overline{\gamma} - t) \tag{5.43}$$

Proof. First, adding (5.33c) scaled by weight $\overline{\beta} > 0$ to (5.33d) yields the inequality

$$y + \overline{\beta}x \leq \overline{\beta} \tag{5.44}$$

which is valid for R^0.

Next, using the substitution $u = xt$ yields

$$(\overline{\gamma} - \underline{\gamma})y + \overline{\beta}(\overline{\gamma}x - u) + \frac{\underline{\gamma}y(u - \underline{\gamma}x)}{y + u - \underline{\gamma}x}$$

$$= (\overline{\gamma} - \underline{\gamma})y + \overline{\beta}x(\overline{\gamma} - t) + \frac{\underline{\gamma}yx(t - \underline{\gamma})}{y + x(t - \underline{\gamma})}$$

$$\leq (\overline{\gamma} - \underline{\gamma})y + \overline{\beta}x(\overline{\gamma} - t) + \frac{\underline{\gamma}yx(t - \underline{\gamma})}{-xt + x(t - \underline{\gamma})} \quad \text{(because } y \leq -xt \text{ and } \underline{\gamma}yx(t - \underline{\gamma}) \leq$$

$$= (\overline{\gamma} - \underline{\gamma})y + \overline{\beta}x(\overline{\gamma} - t) - y(t - \underline{\gamma})$$

$$= \overline{\beta}x(\overline{\gamma} - t) + y(\overline{\gamma} - t)$$

$$= (\overline{\beta}x + y)(\overline{\gamma} - t)$$

$$\leq \overline{\beta}(\overline{\gamma} - t) \qquad \text{because } \overline{\gamma} \geq t \text{ and by (5.44)}.$$

$\qquad\square$

To verify that (5.43) defines a convex constraint, we define $v := u - \underline{\gamma}x$ and need to show that the nonlinear function $f : \mathbb{R}_{>0} \times \mathbb{R}_{\geq 0} \to \mathbb{R}$ with

$$f(y, v) := \frac{\gamma y v}{y + v}$$

is convex. With $\gamma < 0$, this reduces to showing that the function

$$g(y, v) := \frac{yv}{y + v}$$

is concave. The python library Sympy [Sym16] for symbolic mathematics computes the Hessian of g as

$$\left(\begin{matrix} \frac{2vy}{(v+y)^3} - \frac{2y}{(v+y)^2} & \frac{2vy}{(v+y)^3} - \frac{v}{(v+y)^2} - \frac{y}{(v+y)^2} + \frac{1}{v+y} \\ \frac{2vy}{(v+y)^3} - \frac{v}{(v+y)^2} - \frac{y}{(v+y)^2} + \frac{1}{v+y} & \frac{2vy}{(v+y)^3} - \frac{2v}{(v+y)^2} \end{matrix} \right)$$

and its Eigenvalues as

$$\lambda_1 = 0$$

$$\lambda_2 = -\frac{2v^2 + 2y^2}{(v + y)^3}.$$

The second Eigenvalue λ_2 is negative since $v, y \geq 0$. The Hessian is therefore negative semidefinite and g is concave.

The conditionally validity of (5.43) does not pose a problem for the convex hull analysis, but it does for computation. The conditional constraint (5.43), even though convex, is hard do fed into a solver. Moreover, we will prefer to use a linear relaxation so that we want to approximate (5.43) by linear inequalities in a cutting plane scheme. As thing are, we can't be sure that gradient inequalities on (5.43) are valid for T. We strive a convex extension for (5.43) for $y \leq 0$. To this end define the function $h : \mathbb{R} \times \mathbb{R}_{\geq 0} \to \mathbb{R}$ as

$$h(y, v) := \begin{cases} 0 & \text{if } y \leq 0 \\ (\overline{\gamma} - \underline{\gamma})y + f(y, v) & \text{if } y > 0. \end{cases}$$

Next, we show that the inequality with f replaced by h is valid for T and convex.

Lemma 5.11: *If $\overline{\beta} > 0$ and $\underline{\gamma} < 0$, then the inequality*

$$\overline{\beta}(\overline{\gamma}x - u) + h(y, u - \underline{\gamma}x) \leq \overline{\beta}(\overline{\gamma} - t) \tag{5.45}$$

valid for T and convex.

Proof. By Theorem 5.10, the inequality is valid for all points in T with $y > 0$. If $y \leq 0$ the inequality is also valid since

$$\overline{\beta}(\overline{\gamma}x - u) + h(y, u, x) = \overline{\beta}(\overline{\gamma}x - u)$$
$$= x\overline{\beta}(\overline{\gamma} - t)$$
$$\leq \overline{\beta}(\overline{\gamma} - t).$$

To show that the constraint is convex, it suffices to show that h is convex. Let $p_i = (y_i, v_i) \in \mathbb{R} \times \mathbb{R}_{\geq 0}$ for $i = 1, 2$ and $\lambda \in (0, 1)$. We need to show that

$$h(\lambda p_1 + (1 - \lambda)p_2) \leq \lambda h(p_1) + (1 - \lambda)h(p_2). \tag{5.46}$$

If both $y_i > 0$ or both $y_i \leq 0$, then there is nothing to be proved as the point correspond to the same branch in the definition of h. Therefore, assume without loss of generality $y_1 \leq 0$ and $y_2 > 0$. First we show that h is nonnegative. For $y \leq 0$ this is clear and for $y > 0$ we have

$$h(u, v) = (\overline{\gamma} - \underline{\gamma})y + \frac{\gamma y v}{y + v}$$
$$= \frac{(\overline{\gamma} - \underline{\gamma})y(y + v) + \underline{\gamma}yv}{y + v}$$
$$= \frac{(\overline{\gamma} - \underline{\gamma})y^2 + \overline{\gamma}yv}{y + v} \geq 0.$$

Furthermore we know $h(p_1) = 0$ since $y_1 \leq 0$. If $h(\lambda p_1 + (1 - \lambda)p_2) = 0$, then (5.46) is always fulfilled. If $h(\lambda p_1 + (1 - \lambda)p_2)$ does not vanish, then denote by $p_3 = (y_3, v_3)$ the point on the line between p_1 and p_2 with $y_3 = 0$. If $v_3 = 0$, then $v_2 = 0$ since $v_1 \geq 0$. In this case f vanishes such that h is linear on the line between p_3 and p_2 and (5.46) is fulfilled. If $v_3 > 0$, then f is also well-defined at p_3 with $f(p_3) = 0$ so that $h(p_3) = (\overline{\gamma} - \underline{\gamma})y_3 + f(p_3) = 0$ and h is convex on the line between p_3 and p_2. Furthermore, exists a $\hat{\lambda}$ such that

$$\lambda p_1 + (1 - \lambda)p_2 = \hat{\lambda}p_3 + (1 - \hat{\lambda})p_2.$$

and since p_3 is closer to p_2 than p_1, it is $\lambda \leq \hat{\lambda}$. Finally, we can show that (5.46) also holds in this case:

$$h(\lambda p_1 + (1 - \lambda)p_2) = h(\hat{\lambda}p_3 + (1 - \hat{\lambda})p_2)$$
$$\leq \hat{\lambda}h(p_3) + (1 - \hat{\lambda})h(p_2)$$
$$= \lambda h(p_1) + (1 - \hat{\lambda})h(p_2)$$
$$\leq \lambda h(p_1) + (1 - \lambda)h(p_2)$$

$\square$

The previous valid inequalities are nonlinear and convex. The remaining ones are linear.

Theorem 5.12: *The inequality*

$$(\overline{\gamma} - \underline{\gamma})y + \underline{\gamma}(\overline{\gamma}x - u) + \overline{\beta}(u - \underline{\gamma}x) \leq \overline{\beta}(t - \underline{\gamma}) \tag{5.47}$$

is valid for T if $\overline{\beta} > 0$.

Proof. First observe that $y + u \leq 0$ and $-u \leq -\underline{\gamma}x$ together imply

$$y \leq -x\underline{\gamma}. \tag{5.48}$$

Next,

$$
\begin{aligned}
(\overline{\gamma} - \underline{\gamma})y &= (t - \underline{\gamma})y + (\overline{\gamma} - t)y \\
&\leq (t - \underline{\gamma})\overline{\beta}z + (\overline{\gamma} - t)y && \text{because } y \leq \overline{\beta}z \text{ and } t - \underline{\gamma} \geq 0 \\
&\leq (t - \underline{\gamma})\overline{\beta}z + (\overline{\gamma} - t)(-x\underline{\gamma}) && \text{by (5.48) and } \overline{\gamma} - t \geq 0
\end{aligned}
$$

and thus

$$(\overline{\gamma} - \underline{\gamma})y - (t - \underline{\gamma})\overline{\beta}z + (\overline{\gamma} - t)(x\underline{\gamma}) \leq 0. \tag{5.49}$$

Then, multiply the inequality $z + x \leq 1$ on both sides by $\overline{\beta}(t - \underline{\gamma}) \geq 0$ to yield:

$$(t - \underline{\gamma})\overline{\beta}z + (t - \underline{\gamma})\overline{\beta}x \leq (t - \underline{\gamma})\overline{\beta} \tag{5.50}$$

Adding (5.49) and (5.50) yields:

$$(\overline{\gamma} - \underline{\gamma})y + (\overline{\gamma} - t)x\underline{\gamma} + (t - \underline{\gamma})\overline{\beta}x \leq (t - \underline{\gamma})\overline{\beta} \tag{5.51}$$

Finally, substituting $u = xt$ from (5.33a) yields (5.47). □

We next show that if $\overline{\gamma} > 0$, then (5.47) is redundant. Assuming $\overline{\gamma} > 0$, then scaling the inequality $-u + \underline{\gamma}x \leq 0$ by $\overline{\gamma} > 0$ and combining that with the valid inequality $(\overline{\gamma} - \underline{\gamma})y + (\overline{\gamma} - \underline{\gamma})u \leq 0$ and yields

$$(\overline{\gamma} - \underline{\gamma})y - \underline{\gamma}u + \underline{\gamma}\overline{\gamma}x \leq 0. \tag{5.52}$$

But, also since $u - \underline{\gamma}x \leq t - \underline{\gamma}$,

$$\overline{\beta}(u - \underline{\gamma}x) \leq \overline{\beta}(t - \underline{\gamma}).$$

Combining this with (5.52) implies (5.47).

The next theorem presents the last valid inequality for T in this section.

Theorem 5.13: *Inequality*

$$(\gamma - \underline{\beta})(\overline{\gamma}x - u) \leq -\underline{\beta}(\overline{\gamma} - t) \tag{5.53}$$

is valid for T if $\underline{\beta} < 0$.

Proof. Aggregate inequality (5.33c) with weight $-\underline{\beta} > 0$ yields

$$-\underline{\beta}(z + x) \leq -\underline{\beta}. \tag{5.54}$$

Furthermore, using $y \geq \underline{\beta}z$, $\underline{\beta} < 0$, and (5.48), yields $-\underline{\beta}z + x\underline{\gamma} \leq 0$, which combined with (5.54) yields

$$(\underline{\gamma} - \underline{\beta})x \leq -\underline{\beta}.$$

Multiplying both sides of this inequality by $\overline{\gamma} - t \geq 0$ yields

$$(\underline{\gamma} - \underline{\beta})x(\overline{\gamma} - t) \leq -\underline{\beta}(\overline{\gamma} - t).$$

Substituting $xt = u$ yields (5.53). □

If $\underline{\gamma} < 0$, then $(\underline{\gamma} - \underline{\beta})(\overline{\gamma}x - u) \leq -\underline{\beta}(\overline{\gamma}x - 0) \leq -\underline{\beta}(\overline{\gamma} - t)$ and so (5.53) is redundant.

5.3.4 Convex hull analysis

We next demonstrate that the set R^0 combined with certain subsets of the new valid inequalities, depending on the sign of $\underline{\gamma}$ and $\overline{\gamma}$, are sufficient to define the convex hull of T. Let us first restate the relevant inequalities for convenience:

$$(u - \underline{\beta}x)(u - \underline{\gamma}x) \leq -\underline{\beta}x(t - \underline{\gamma}) \tag{5.40}$$

$$(\overline{\gamma} - \underline{\gamma})y + \overline{\beta}(\overline{\gamma}x - u) + \frac{\underline{\gamma}y(u - \underline{\gamma}x)}{y + u - \underline{\gamma}x} \leq \overline{\beta}(\overline{\gamma} - t) \quad \text{if } y > 0. \tag{5.43}$$

$$(\overline{\gamma} - \underline{\gamma})y + \underline{\gamma}(\overline{\gamma}x - u) + \overline{\beta}(u - \underline{\gamma}x) \leq \overline{\beta}(t - \underline{\gamma}) \tag{5.47}$$

$$(\underline{\gamma} - \underline{\beta})(\overline{\gamma}x - u) \leq -\underline{\beta}(\overline{\gamma} - t). \tag{5.53}$$

Next, we define the sets which include the non-redundant valid inequalities for different signs of $\underline{\gamma}$ and $\overline{\gamma}$:

$$R^1 = \{(x, u, y, z, t) \in R^0 : \text{(5.40) and (5.43)}\},$$
$$R^2 = \{(x, u, y, z, t) \in R^0 : \text{(5.43) and (5.47)}\},$$
$$R^3 = \{(x, u, y, z, t) \in R^0 : \text{(5.40) and (5.53)}\}.$$

Result	Used in proof of	Stated on
Lemma 5.18	Theorems 5.14 to 5.16	Page 116
Lemma 5.21	Theorem 5.14	Page 118
Lemma 5.22	Theorem 5.14	Page 121
Lemma 5.23	Theorem 5.15	Page 124
Lemma 5.24	Theorem 5.15	Page 124
Lemma 5.26	Theorem 5.16	Page 127
Lemma 5.27	Theorem 5.16	Page 129

Table 5.2: Overview over lemmas used in the proofs of Theorems 5.14 to 5.16

The following theorems show that R^1, R^2, and R^3 describe the convex full of T in different cases. Since all inequalities that define the sets are valid for T and convex, R^i are convex relaxations for T and we know $\text{conv}(T) \subset R^i$ for $i = 1, \ldots, 4$. To show that a relaxation defines $\text{conv}(T)$ in a particular case, we show that every extreme point of the relaxation satisfies the non-convex constraint $u = xt$ even though this equation is not enforced in the relaxation. The theorems are stated in the following and are proved using lemmas that are stated and proved thereafter. Table 5.2 gives on overview which of the lemmas is used in the proof of which theorem and on which page the lemmas are found. We believe this gives the reader the best overview over the results and the structure of the proofs.

Theorem 5.14: *Assume $\underline{\gamma} < 0 < \overline{\gamma}$ and $\underline{\beta} < 0 < \overline{\beta}$. Then $\text{conv}(T) = R^1$.*

Proof. Being a bounded convex set, R^1 is completely characterized by its extreme points. We prove that every extreme point of R^1 is in T, i.e., fulfills the equation $u = xt$. Every point in R^1 can thus be represented as a convex combination of points in T and $R^1 = \text{conv}(T)$ is proved.

It remains to show that $u = xt$ for all $p = (x, u, y, z, t) \in \text{ext}(R^1)$. Lemma 5.18 show that this is the case of x or t are at its bounds, i.e., if $x \in \{0, 1\}$ or $t \in \{\underline{\gamma}, \overline{\gamma}\}$. Under the condition $0 < x < 1$ and $\underline{\gamma} < t < \overline{\gamma}$, Lemmas 5.21 and 5.22 show that points with $u < xt$ and $u > xt$, respectively, cannot be extreme. □

Theorem 5.15: *Assume $\underline{\gamma} < \overline{\gamma} < 0$ and $\underline{\beta} < 0 < \overline{\beta}$. Then $\text{conv}(T) = R^2$.*

Proof. Based on the same argument as in the proof of Theorem 5.14, we show that that $u = xt$ for all $p = (x, u, y, z, t) \in \text{ext}(R^2)$. Lemma 5.18 shows that for $x \in \{0, 1\}$ or $t \in \{\underline{\gamma}, \overline{\gamma}\}$. For $0 < x < 1$ and $\underline{\gamma} < t < \overline{\gamma}$ and under the assumptions of this theorem Lemmas 5.23 and 5.24 show that points with $u < xt$ and $u > xt$, respectively, cannot be extreme. □

Theorem 5.16: *Assume* $0 < \underline{\gamma} < \overline{\gamma}$ *and* $\underline{\beta} < 0 < \overline{\beta}$. *Then* $\mathrm{conv}(T) = R^3$.

Proof. The proof is analogous to the proofs of Theorems 5.14 and 5.15, but in this case Lemmas 5.26 and 5.27 show that points with $u < xt$ and $u > xt$, respectively, cannot be extreme. $\qquad\square$

Preliminary Results

In the following, for different assumptions on the sign of $\underline{\gamma}$ and $\overline{\gamma}$, we demonstrate that if $p = (x, u, y, z, t)$ has either $u > xt$ or $u < xt$, then p is not an extreme point of $\mathrm{conv}(T)$. This will be accomplished by considering different cases and in each case, we provide two distinct points which depend on a parameter $\epsilon > 0$, denoted $p_i^\epsilon, i = 1, 2$, which satisfy $p = (1/2)p_1^\epsilon + (1/2)p_2^\epsilon$. Furthermore, the points p_i^ϵ are defined such that $p_i^\epsilon \to p$ as $\epsilon \to 0$. The points are then shown to be in the given relaxation for $\epsilon > 0$ small enough, providing a proof that p is not an extreme point of the relaxation. To show the points are in a given relaxation for $\epsilon > 0$ small enough, for each inequality defining the relaxation we either directly show the points satisfy the inequality, or else we show that the point p satisfies the inequality with strict inequality. In the latter case, the following lemma ensures that both points p_1^ϵ and p_2^ϵ satisfy the constraint if ϵ is small enough.

Proposition 5.17: *Let* $p^\epsilon : \mathbb{R}_+ \to \mathbb{R}^n$ *with* $\lim_{\epsilon \to 0} p^\epsilon = p$ *for some* $p \in \mathbb{R}^n$. *Suppose* $ap < b$ *for* $a \in \mathbb{R}^n$, $b \in \mathbb{R}$. *Then there exists an* $\hat{\epsilon} > 0$ *such that*

$$ap^\epsilon < b \qquad \text{for all } \epsilon < \hat{\epsilon}.$$

Proof. Follows directly from continuity of the function $f(x) = ax - b$. $\qquad\square$

Throughout this section, for $\epsilon > 0$, we use the notation:

$$\alpha_1^\epsilon = 1 - \epsilon, \ \alpha_2^\epsilon = 1 + \epsilon \qquad \text{and} \qquad \delta_1^\epsilon = \epsilon, \ \delta_2^\epsilon = -\epsilon.$$

Obviously, $\lim_{\epsilon \to 0} \alpha_i^\epsilon = 1$ and $\lim_{\epsilon \to 0} \delta_i^\epsilon = 0$ for $i \in \{1, 2\}$.

The series of Lemmas that prove Theorems 5.14 to 5.16 is started by Lemma 5.18 which applies to all cases and tells us that points on the boundaries of the domains of x and t fulfill $u = xt$.

Lemma 5.18: *Let* $p = (x, u, y, z, t) \in R^0$. *If* $x = 0$, $x = 1$, $t = \underline{\gamma}$, *or* $t = \overline{\gamma}$, *then* $u = xt$.

Proof. This follows since (5.34)–(5.37) are the McCormick inequalities for relaxing the constraint $u = xt$ over $x \in [0, 1]$ and $t \in [\underline{\gamma}, \overline{\gamma}]$, and it is known (e.g., [AF83]) that if either of the variables are at its bound, then the McCormick inequalities ensure that $u = xt$. $\qquad\square$

As Lemma 5.18 applies to all the cases we assume from now on that $0 < x < 1$ and $\underline{\gamma} < t < \overline{\gamma}$. We use the following two propositions in several places in this section.

Proposition 5.19: *Suppose $\underline{\beta} < 0$. Let $p = (x, u, y, z, t) \in R^0$ with $0 < x < 1$ and $\underline{\gamma} < t < \overline{\gamma}$.*

1. *If $u < xt$, then p satisfies (5.35), (5.36), and (5.40) with strict inequality.*

2. *If $u > xt$, then p satisfies (5.34) and (5.37) with strict inequality.*

Proof. 1. Suppose $u < xt$. Then, $\overline{\gamma}x - u > \overline{\gamma}x - xt = x(\overline{\gamma} - t) > 0$, and so p satisfies (5.35) with strict inequality. Next,

$$u - \underline{\gamma}x < xt - \underline{\gamma}x = x(t - \underline{\gamma}) < t - \underline{\gamma} \tag{5.55}$$

as $x < 1$ and $t > \underline{\gamma}$, and so (5.36) is satisfied by p with strict inequality. To show that (5.40) is satisfied strictly, we aggregate (5.33b) with weight 1, (5.33c) with weight $-\beta$, and (5.33e) with weight 1 and get

$$u - \underline{\beta}x \le -\underline{\beta}. \tag{5.56}$$

As $u - \underline{\gamma}x \ge 0$,

$$(u - \underline{\beta}x)(u - \underline{\gamma}x) \le -\underline{\beta}(u - \underline{\gamma}x) < -\underline{\beta}x(t - \underline{\gamma})$$

where the last inequality follows from (5.55) and $\underline{\beta} < 0$, and thus (5.40) is satisfied by p with strict inequality.

2. Now suppose $u > xt$. Then, $u - x\underline{\gamma} > xt - x\underline{\gamma} = x(t - \underline{\gamma}) > 0$ and so p satisfies (5.34) with strict inequality. Next,

$$\overline{\gamma}x - u < \overline{\gamma}x - xt = x(\overline{\gamma} - t) < \overline{\gamma} - t \tag{5.57}$$

as $x < 1$ and $t < \overline{\gamma}$, and so (5.37) is satisfied with strict inequality. $\qquad\square$

Proposition 5.20: *Suppose $\underline{\beta} < 0$ and $\underline{\gamma} < 0$. Let $p = (x, u, y, z, t) \in R^0$ with $0 < x < 1$ and $\underline{\gamma} < t < \overline{\gamma}$. If $u > xt$ and $y > 0$, then p satisfies (5.43) with strict inequality.*

Proof. Then, as $\underline{\gamma}y < 0$, we have

$$\underline{\gamma}y(u - \underline{\gamma}x) < \underline{\gamma}y(xt - \underline{\gamma}x) = \underline{\gamma}yx(t - \underline{\gamma}).$$

Thus, as $y + u - \underline{\gamma}x > 0$,

$$\frac{\underline{\gamma}y(u - \underline{\gamma}x)}{y + u - \underline{\gamma}x} < \frac{\underline{\gamma}yx(t - \underline{\gamma})}{y + u - \underline{\gamma}x} \le \frac{\underline{\gamma}yx(t - \underline{\gamma})}{-\underline{\gamma}x} = -y(t - \underline{\gamma})$$

where the last inequality follows from $y + u \leq 0$ and $\gamma y x (t - \gamma) < 0$. Thus,

$$(\overline{\gamma} - \underline{\gamma})y + \overline{\beta}(\underline{\gamma}x - u) + \frac{\gamma y(u - \gamma x)}{y + u - \gamma x} < (\overline{\gamma} - \underline{\gamma})y + \overline{\beta}(\underline{\gamma}x - u) - y(t - \underline{\gamma})$$

$$< y(\overline{\gamma} - t) + \overline{\beta}x(\overline{\gamma} - t)$$

$$= (y + \overline{\beta}x)(\overline{\gamma} - t) \leq \overline{\beta}(\overline{\gamma} - t)$$

where the last inequality follows from $\overline{\gamma} - t > 0$ and the fact that aggregating (5.33c) with weight $\overline{\beta}$ and (5.33d) yields $y + \overline{\beta}x \leq \overline{\beta}$. $\qquad\square$

Proof of Theorem 5.14

We now state and prove the two main lemmas that support the proof of Theorem 5.14.

Lemma 5.21: *Suppose $\underline{\beta} < 0 < \overline{\beta}$. Let $p = (x, u, y, z, t) \in R^1$ with $0 < x < 1$ and $\underline{\gamma} < t < \overline{\gamma}$. If $u < xt$, then p is not an extreme point of R^1.*

Proof. We consider four cases: (a) $y + u < 0$, (b) $z + x < 1$, (c) $\underline{\beta}z - y < 0$ and $y - \overline{\beta}z < 0$, and (d) $z + x = 1$, $y + u = 0$, and either $\underline{\beta}z - y = 0$ or $y - \overline{\beta}z = 0$. In each of them we define a series of points $p_i^\epsilon = (x_i^\epsilon, u_i^\epsilon, y_i^\epsilon, z_i^\epsilon, t_i^\epsilon)$ for $i \in \{1, 2\}$ that depends on $\epsilon > 0$ with $p = 0.5(p_1^\epsilon + p_2^\epsilon)$ and which satisfy $\lim_{\epsilon \to 0} p_i^\epsilon = p$. We then show that both p_i^ϵ are in R^1 and thus p is not an extreme point of R^1. To show $p_i^\epsilon \in R^1$, we need to ensure that it satisfies all inequalities defining R^1. For those inequalities that satisfied strictly at p, Proposition 5.17 ensures that this is the case. For the remaining inequalities, we show it directly.

By Proposition 5.19, $u < xt$ implies that p satisfies (5.35), (5.36), and (5.40) with strict inequality. It remains to show that the points p_i^ϵ satisfy (5.33b)–(5.33e), (5.34), (5.37), and (5.43) for $\epsilon > 0$ small enough. Note that $z \geq 0$ is implied by (5.33d) and (5.33e) and does not have to be proved explicitly.

Case (a): $y + u < 0$. For $\epsilon > 0$, define $p_i^\epsilon = (x_i^\epsilon, u_i^\epsilon, y_i^\epsilon, z_i^\epsilon, t_i^\epsilon)$ where

$$x_i^\epsilon := (1 - \alpha_i^\epsilon) + \alpha_i^\epsilon x \qquad u_i^\epsilon := \underline{\gamma}(1 - \alpha_i^\epsilon) + \alpha_i^\epsilon u \qquad y_i^\epsilon := \alpha_i^\epsilon y$$

$$z_i^\epsilon := \alpha_i^\epsilon z \qquad\qquad\qquad t_i^\epsilon := (1 - \alpha_i^\epsilon)\underline{\gamma} + \alpha_i^\epsilon t$$

for $i = 1, 2$. Since α_i^ϵ converge to 1, it is clear that p_i^ϵ converges to p and Proposition 5.17 can be applied.

In the following we will go one-by-one through the inequalities to be checked.

(5.33b) Satisfied strictly by p by the assumption of this case.

(5.33c)–(5.33e) Easily checked directly.

(5.34) Follows from

$$u_i^\epsilon - \underline{\gamma} x_i^\epsilon = \underline{\gamma}(1 - \alpha_i^\epsilon) + \alpha_i^\epsilon u - \underline{\gamma}((1 - \alpha_i^\epsilon) + \alpha_i^\epsilon x) = \alpha_i^\epsilon (u - \underline{\gamma} x) \geq 0.$$

(5.37) Follows from

$$\begin{aligned}
\overline{\gamma} x_i^\epsilon - u_i^\epsilon &= \overline{\gamma}((1 - \alpha_i^\epsilon) + \alpha_i^\epsilon x) - \underline{\gamma}(1 - \alpha_i^\epsilon) - \alpha_i^\epsilon u \\
&= (1 - \alpha_i^\epsilon)(\overline{\gamma} - \underline{\gamma}) + \alpha_i^\epsilon(\overline{\gamma} x - u) \\
&\leq (1 - \alpha_i^\epsilon)(\overline{\gamma} - \underline{\gamma}) + \alpha_i^\epsilon(\overline{\gamma} - t) = \overline{\gamma} - (1 - \alpha_i^\epsilon)\underline{\gamma} - \alpha_i^\epsilon t = \overline{\gamma} - t_i^\epsilon.
\end{aligned}$$

(5.43) If $y > 0$, then also $y_i^\epsilon > 0$, and using $u_i^\epsilon - \underline{\gamma} x_i^\epsilon = \alpha_i^\epsilon(u - \underline{\gamma} x)$ and $\overline{\gamma} x_i^\epsilon - u_i^\epsilon = (1 - \alpha_i^\epsilon)(\overline{\gamma} - \underline{\gamma}) + \alpha_i^\epsilon(\overline{\gamma} x - u)$, the left-hand-side of (5.43) evaluated at p_i^ϵ is equal to:

$$\alpha_i^\epsilon\left((\underline{\gamma} - \overline{\gamma})y + \overline{\beta}(\overline{\gamma} x - u) + \frac{\underline{\gamma} y(u - \underline{\gamma} x)}{y + u + \underline{\gamma} x}\right) + \overline{\beta}(1 - \alpha_i^\epsilon)(\underline{\gamma} - \overline{\gamma})$$

$$\leq \alpha_i^\epsilon \overline{\beta}(\overline{\gamma} - t) + \overline{\beta}(1 - \alpha_i^\epsilon)(\underline{\gamma} - \overline{\gamma}) = \overline{\beta}(\overline{\gamma} - \alpha_i^\epsilon t - (1 - \alpha_i^\epsilon)\underline{\gamma}) = \overline{\beta}(\overline{\gamma} - t_i^\epsilon)$$

and hence (5.43) is satisfied by p_i^ϵ for $i = 1, 2$ and any $\epsilon \in (0, 1)$ when $y > 0$. On the other hand, if $y \leq 0$, then $y_i^\epsilon \leq 0$, and p_i^ϵ is not required to satisfy (5.43) for $i = 1, 2$.

Case (b): $z + x < 1$. For $\epsilon > 0$, define $p_i^\epsilon = (x_i^\epsilon, u_i^\epsilon, y_i^\epsilon, z_i^\epsilon, t_i^\epsilon)$ where

$$\begin{aligned}
x_i^\epsilon &:= \alpha_i^\epsilon x & u_i^\epsilon &:= \alpha_i^\epsilon u & y_i^\epsilon &:= \alpha_i^\epsilon y \\
z_i^\epsilon &:= \alpha_i^\epsilon z & t_i^\epsilon &:= \alpha_i^\epsilon t + (1 - \alpha_i^\epsilon)\overline{\gamma}
\end{aligned}$$

for $i = 1, 2$.

(5.33b) Easily checked directly.

(5.33c) Satisfied strictly by p by the assumption of this case.

(5.33d), (5.33e), and (5.34) Easily checked directly.

(5.37) Follows from

$$\overline{\gamma} x_i^\epsilon - u_i^\epsilon = \alpha_i^\epsilon(\overline{\gamma} x - u) \leq \alpha_i^\epsilon(\overline{\gamma} - t) = \alpha_i^\epsilon \overline{\gamma} - \alpha_i^\epsilon t = \alpha_i^\epsilon \overline{\gamma} - (1 - \alpha_i^\epsilon)\overline{\gamma} - t_i^\epsilon = \overline{\gamma} - t_i^\epsilon.$$

(5.43) If $y > 0$, then also $y_i^\epsilon > 0$, and the left-hand-side of (5.43) evaluated at p_i^ϵ is equal to:

$$\begin{aligned}
\alpha_i^\epsilon\left((\underline{\gamma} - \overline{\gamma})y + \overline{\beta}(\overline{\gamma} x - u) + \frac{\underline{\gamma} y(u - \underline{\gamma} x)}{y + u + \underline{\gamma} x}\right) &\leq \alpha_i^\epsilon \overline{\beta}(\overline{\gamma} - t) \\
&= \overline{\beta}(\alpha_i^\epsilon \overline{\gamma} - t_i^\epsilon + (1 - \alpha_i^\epsilon)\overline{\gamma}) \\
&= \overline{\beta}(\overline{\gamma} - t_i^\epsilon)
\end{aligned}$$

and hence (5.43) is satisfied by p_i^ϵ for $i = 1, 2$ and any $\epsilon \in (0, 1)$ when $y > 0$. On the other hand, if $y \leq 0$, then $y_i^\epsilon \leq 0$, and p_i^ϵ is not required to satisfy (5.43) for $i = 1, 2$.

Case (c): $\underline{\beta} z - y < 0$ **and** $y - \overline{\beta} z < 0$. For $\epsilon > 0$, define $p_i^\epsilon = (x_i^\epsilon, u_i^\epsilon, y_i^\epsilon, z_i^\epsilon, t_i^\epsilon)$ where

$$x_i^\epsilon := \alpha_i^\epsilon x \qquad\qquad u_i^\epsilon := \alpha_i^\epsilon u \qquad\qquad y_i^\epsilon := \alpha_i^\epsilon y$$
$$z_i^\epsilon := (1 - \alpha_i^\epsilon) + \alpha_i^\epsilon z \qquad t_i^\epsilon := \alpha_i^\epsilon t + (1 - \alpha_i^\epsilon)\overline{\gamma}$$

for $i = 1, 2$.
(5.33b) Easily checked directly.
(5.33c) Follows from

$$z_i^\epsilon + x_i^\epsilon = (1 - \alpha_i^\epsilon) + \alpha_i^\epsilon z + \alpha_i^\epsilon x = (1 - \alpha_i^\epsilon) + \alpha_i^\epsilon (z + x) \leq 1.$$

(5.33d), (5.33e) Satisfied strictly by p by the assumption of this case.
(5.34) Easily checked directly.
(5.37), (5.43) As the definitions of $t_i^\epsilon, y_i^\epsilon, x_i^\epsilon$, and u_i^ϵ are the same as in Case (b), it follows from the arguments in that case that p_i^ϵ satisfies (5.37) and (5.43) for $i = 1, 2$ and any $\epsilon \in (0, 1)$.

Case (d): $z + x = 1$, $y + u = 0$, **and either** $\underline{\beta} z - y = 0$ **or** $y - \overline{\beta} z = 0$. For $\epsilon > 0$, define $p_i^\epsilon = (x_i^\epsilon, u_i^\epsilon, y_i^\epsilon, z_i^\epsilon, t_i^\epsilon)$ where

$$x_i^\epsilon := x - \delta_i^\epsilon \qquad\qquad u_i^\epsilon := u - \underline{\beta}\delta_i^\epsilon \qquad\qquad y_i^\epsilon := y + \underline{\beta}\delta_i^\epsilon$$
$$z_i^\epsilon := z + \delta_i^\epsilon \qquad\qquad t_i^\epsilon := t + \delta_i^\epsilon(\overline{\gamma} - \underline{\beta})$$

for $i = 1, 2$.
(5.33b), (5.33c) Easily checked directly.
(5.33d) We show that when $z + x = 1$ and $y + u = 0$, then $y - \overline{\beta} z < 0$. Indeed, if $y - \overline{\beta} z = 0$, then as $z + x = 1$, it follows that

$$y + \overline{\beta} x = \overline{\beta} \tag{5.58}$$

Then, using $y = \overline{\beta} z > 0$, and evaluating p in the left-hand-side of (5.43) yields

$$(\overline{\gamma} - \underline{\gamma})y + \overline{\beta}(\overline{\gamma}x - u) + \frac{\underline{\gamma}y(u - \underline{\gamma}x)}{y + u - \underline{\gamma}x}$$

$$= (\overline{\gamma} - \underline{\gamma})y + \overline{\beta}(\overline{\gamma}x - u) + \frac{\underline{\gamma}y(u - \underline{\gamma}x)}{-\underline{\gamma}x} \qquad \text{since } y + u = 0$$

$$= \overline{\gamma}(y + \overline{\beta}x) - u(\overline{\beta} + y/x)$$

$$> \overline{\gamma}(y + \overline{\beta}x) - xt(\overline{\beta} + y/x) \qquad \text{since } u < xt \text{ and } \overline{\beta} + y/x > 0$$

$$= (\overline{\gamma} - t)(y + \overline{\beta}x) = (\overline{\gamma} - t)\overline{\beta} \qquad \text{by (5.58).}$$

Thus, p violates (5.43) and hence p fulfills (5.33d) with strict inequality. Furthermore, due to the assumptions of this case, we can assume $\underline{\beta}z - y = 0$.

(5.33e) Easily checked directly.

(5.34) As $\underline{\beta}z - y = 0$, $z > 0$ and $s > 0$, p satisfies (5.34) with strict inequality:

$$u - \underline{\gamma}x = -y - \underline{\gamma}x = -\underline{\beta}z - \underline{\gamma}x > 0.$$

(5.37) Follows from

$$\overline{\gamma}x_i^\epsilon - u_i^\epsilon = \overline{\gamma}x - \overline{\gamma}\delta_i^\epsilon - u + \underline{\beta}\delta_i^\epsilon \leq \overline{\gamma} - t + \delta_i^\epsilon(\underline{\beta} - \overline{\gamma}) = \overline{\gamma} - t_i^\epsilon.$$

(5.43) Because $z > 0$ and $y = \underline{\beta}z < 0$, it follows that $y_i^\epsilon < 0$ and p_i^ϵ is not subject to (5.43).

$\square$

Lemma 5.22: *Suppose $\overline{\gamma} > 0$ and $\underline{\beta} < 0 < \overline{\beta}$. Let $p = (x, u, y, z, t) \in R^1$ with $0 < x < 1$ and $\underline{\gamma} < t < \overline{\gamma}$. If $u > xt$ and either $y \leq 0$ or p satisfies (5.43) with strict inequality, then p is not an extreme point of R^1.*

We first comment that the assumption that either $y \leq 0$ or p satisfies (5.43) with strict inequality follows from the assumption $u > xt$ and Proposition 5.20 when $\underline{\gamma} < 0$. However, we state the assumption in this way in order to make the applicability of this proposition clear for a later case when $\underline{\gamma} > 0$.

Proof. This proof has the same structure as the proof of Lemma 5.21. By Proposition 5.19, $u > xt$ implies that p satisfies (5.34) and (5.37) with strict inequality. Also, by assumption, if $y > 0$, then p satisfies (5.43) with strict inequality. It remains to show that the points p_i^ϵ satisfy (5.33b)–(5.33e), (5.35), (5.36), and (5.40) for ϵ small enough. We consider four cases.

Case (a): $y + u < 0$ **and** $z + x = 1$. Note that $z + x = 1$ and $x < 1$ implies that $z > 0$. Thus, either (5.33d) or (5.33e) is satisfied strictly by p. If $y - \overline{\beta}z < 0$, define $y_i^\epsilon := (1 - \alpha_i^\epsilon)\underline{\beta} + \alpha_i^\epsilon y$, and otherwise, if $\underline{\beta}z - y < 0$, define $y_i^\epsilon := (1 - \alpha_i^\epsilon)\overline{\beta} + \alpha_i y$ for $\epsilon > 0$. Then, for $\epsilon > 0$, define $p_i^\epsilon = (x_i^\epsilon, u_i^\epsilon, y_i^\epsilon, z_i^\epsilon, t_i^\epsilon)$ where

$$x_i^\epsilon = \alpha_i^\epsilon x \qquad\qquad u_i^\epsilon = \alpha_i^\epsilon u$$
$$z_i^\epsilon = (1 - \alpha_i^\epsilon) + \alpha_i^\epsilon z \qquad\qquad t_i^\epsilon = (1 - \alpha_i^\epsilon)\underline{\gamma} + \alpha_i^\epsilon t$$

for $i = 1, 2$.

(5.33b) Satisfied strictly by p by the assumption of this case.

(5.33c) Easily checked directly.

(5.33d), (5.33e) Recall that either (5.33d) or (5.33e) is satisfied strictly by p. In the case $y - \overline{\beta}z < 0$, i.e., (5.33d) is satisfied strictly, we only need to check p_i^ϵ satisfies (5.33e):

$$\underline{\beta}z_i^\epsilon - y_i^\epsilon = \underline{\beta}((1 - \alpha_i^\epsilon) + \alpha_i^\epsilon z) - ((1 - \alpha_i^\epsilon)\underline{\beta} + \alpha_i^\epsilon y) = \alpha_i^\epsilon(\underline{\beta}z - y) \leq 0.$$

On the other hand, if $\underline{\beta}z - y < 0$, i.e., (5.33e) is satisfied strictly, then

$$y_i^\epsilon - \overline{\beta}z_i^\epsilon = \overline{\beta}(1 - \alpha_i^\epsilon) + \alpha_i^\epsilon y - \overline{\beta}((1 - \alpha_i^\epsilon) + \alpha_i^\epsilon z) = \alpha_i^\epsilon(y - \overline{\beta}z) \leq 0.$$

(5.35) Easily checked directly.

(5.36) Shown directly by

$$u_i^\epsilon - \underline{\gamma}x_i^\epsilon = \alpha_i^\epsilon(u - \underline{\gamma}x) \leq \alpha_i^\epsilon(t - \underline{\gamma}) = t_i^\epsilon - (1 - \alpha_i^\epsilon)\underline{\gamma} - \alpha_i^\epsilon\underline{\gamma} = t_i^\epsilon - \underline{\gamma}.$$

(5.40) Shown directly by

$$\begin{aligned}
(u_i^\epsilon - \underline{\beta}x_i^\epsilon)(u_i^\epsilon - \underline{\gamma}x_i^\epsilon) &= (\alpha_i^\epsilon)^2(u - \underline{\beta}x)(u - \underline{\gamma}x) \\
&\leq (\alpha_i^\epsilon)^2(-\underline{\beta})x(t - \underline{\gamma}) \\
&= -\underline{\beta}(x_i^\epsilon\alpha_i^\epsilon(t - \underline{\gamma})) = -\underline{\beta}x_i^\epsilon(t_i^\epsilon - \underline{\gamma}).
\end{aligned}$$

Case (b): $z + x < 1$. For $\epsilon > 0$, define $p_i^\epsilon = (x_i^\epsilon, u_i^\epsilon, y_i^\epsilon, z_i^\epsilon, t_i^\epsilon)$ where

$$x_i^\epsilon := \alpha_i^\epsilon x \qquad u_i^\epsilon := \alpha_i^\epsilon u \qquad\qquad y_i^\epsilon := \alpha_i^\epsilon y$$
$$z_i^\epsilon := \alpha_i^\epsilon z \qquad t_i^\epsilon := (1 - \alpha_i^\epsilon)\underline{\gamma} + \alpha_i^\epsilon t$$

for $i = 1, 2$. It is clear that (5.33b), (5.33d), (5.33e), (5.34), and (5.35) are satisfied by p_i^ϵ for $i = 1, 2$. By the assumption of this case (5.33c) is strictly satisfied by p. The remaining inequalities (5.36) and (5.40) depend only on the variables x, u, and t, and the definitions of $u_i^\epsilon, x_i^\epsilon$, and t_i^ϵ are the same as in Case (a).

Case (c): $y + u = 0$, $z + x = 1$, **and** $y - \bar{\beta}z < 0$. For $\epsilon > 0$, define $p_i^\epsilon = (x_i^\epsilon, u_i^\epsilon, y_i^\epsilon, z_i^\epsilon, t_i^\epsilon)$ where

$$x_i^\epsilon := \alpha_i^\epsilon x \qquad\qquad u_i^\epsilon := \alpha_i^\epsilon u \qquad\qquad y_i^\epsilon := \alpha_i^\epsilon y$$
$$z_i^\epsilon := (1 - \alpha_i^\epsilon) + \alpha_i^\epsilon z \qquad\qquad t_i^\epsilon := (1 - \alpha_i^\epsilon)\gamma + \alpha_i^\epsilon t$$

for $i = 1, 2$.

(5.33b), (5.33c) Easily checked directly.

(5.33d) Satisfied strictly by p by the assumption of this case.

(5.33e) We show that (5.33e) is satisfied strictly by p. Indeed, if $\underline{\beta}z - y = 0$, then the other equations for this case imply that $u - \underline{\beta}x = -\underline{\beta}$. Then, evaluating p in the left-hand-side of (5.40) yields:

$$(u - \underline{\beta}x)(u - \underline{\gamma}x) = -\underline{\beta}(u - \underline{\gamma}x) > -\underline{\beta}(xt - \underline{\gamma}x) = -\underline{\beta}x(t - \underline{\gamma})$$

and so p violates (5.40).

(5.35) Easily checked directly.

(5.36), (5.40) As the definitions of x_i^ϵ, u_i^ϵ, and t_i^ϵ are the same as in Case (a), the arguments in that case imply p_i^ϵ satisfies (5.36) and (5.40) for $\epsilon \in (0, 1)$.

Case (d): $y + u = 0$, $z + x = 1$, **and** $y - \bar{\beta}z = 0$. For $\epsilon > 0$, define $p_i^\epsilon = (x_i^\epsilon, u_i^\epsilon, y_i^\epsilon, z_i^\epsilon, t_i^\epsilon)$ where

$$x_i^\epsilon := x - \delta_i \qquad\qquad u_i^\epsilon := u - \delta_i\bar{\beta} \qquad\qquad y_i^\epsilon := y + \delta_i\bar{\beta}$$
$$z_i^\epsilon := z + \delta_i \qquad\qquad t_i^\epsilon := t - \delta_i(\bar{\beta} - \underline{\gamma})$$

for $i = 1, 2$.

(5.33b), (5.33c), (5.33d) Easily checked directly.

(5.33e) (5.33e) is satisfied strictly by p by the same argument as in the previous case.

(5.35) We show that $\bar{\gamma}x - u > 0$, i.e., (5.35) is satisfied strictly by p. Indeed, the three equations in this case imply that $\bar{\beta}x - u = \bar{\beta}$. Thus,

$$\bar{\gamma}x - u = \bar{\gamma}x - \bar{\beta}x + \bar{\beta} = \bar{\gamma}x + (1 - x)\bar{\beta} > 0.$$

(5.36) Shown directly by

$$u_i^\epsilon - \underline{\gamma}x_i^\epsilon = u - \delta_i^\epsilon\bar{\beta} - \underline{\gamma}(x - \delta_i^\epsilon) = u - \underline{\gamma}x - \delta_i(\bar{\beta} - \underline{\gamma})$$
$$\leq t - \underline{\gamma} - \delta_i(\bar{\beta} - \underline{\gamma}) = t_i - \underline{\gamma}. \qquad (5.59)$$

(5.40) As $y = \bar{\beta}z$ and $z > 0$, this implies $y > 0$ and in turn $u < 0$. Thus,

$$u - \underline{\beta}x < -\underline{\beta}x$$

and so, for $\epsilon > 0$ small enough, also $u_i^\epsilon - \underline{\beta} x_i^\epsilon < -\underline{\beta} x_i^\epsilon$. Combining this with (5.59) yields

$$(u_i^\epsilon - \underline{\beta} x_i^\epsilon)(u_i^\epsilon - \underline{\gamma} x_i^\epsilon) \le -\underline{\beta} x_i^\epsilon (t_i - \underline{\gamma}).$$

$\square$

Proof of Theorem 5.15

We now state and prove the two main lemmas that support the proof of Theorem 5.15.

Lemma 5.23: *Suppose $\underline{\gamma} < \overline{\gamma} < 0$ and $\underline{\beta} < 0 < \overline{\beta}$. Let $p = (x, u, y, z, t) \in R^2$ with $0 < x < 1$ and $\underline{\gamma} < t < \overline{\gamma}$. If $u < xt$, then p is not an extreme point of R^2.*

Proof. First, we show that p satisfies (5.47) with strict inequality. Observe that the inequality (5.51) is valid for any point in R^2. Thus,

$$(\overline{\gamma} - \underline{\gamma})y + \underline{\gamma}(\overline{\gamma}x - u) + \overline{\beta}(u - \underline{\gamma}x)$$
$$< (\overline{\gamma} - \underline{\gamma})y + \underline{\gamma}(\overline{\gamma}x - xt) + \overline{\beta}(xt - \underline{\gamma}x)$$
$$= (\overline{\gamma} - \underline{\gamma})y + (\overline{\gamma} - t)x\underline{\gamma} + (t - \underline{\gamma})\overline{\beta}x \le (t - \underline{\gamma})\overline{\beta}.$$

When $u < xt$, the inequality (5.47) is satisfied with strict inequality, just as (5.40) is satisfied by strict inequality when $u < xt$ and $\overline{\gamma} > 0$. As the substitution of (5.47) for (5.40) is the only difference between the sets R^2 and R^1, the arguments of Lemma 5.21 apply directly to this case, and we can conclude that if $u < xt$, $0 < x < 1$, and $\underline{\gamma} < t < \overline{\gamma}$, then p is not an extreme point of R^2. $\square$

Lemma 5.24: *Suppose $\underline{\gamma} < \overline{\gamma} < 0$ and $\underline{\beta} < 0 < \overline{\beta}$. Let $p = (x, u, y, z, t) \in R^2$ with $0 < x < 1$ and $\underline{\gamma} < t < \overline{\gamma}$. If $u > xt$, then p is not an extreme point of R^2.*

Proof. This proof has the same structure as the proof of Lemma 5.21. First, by Proposition 5.19, $u > xt$ implies that p satisfies (5.34) and (5.37) with strict inequality, and by Proposition 5.20 if also $y > 0$, then p satisfies (5.43) with strict inequality. Also, as $\overline{\gamma} < 0$, it follows from $u \le \overline{\gamma}x$ and $x > 0$ that $u < 0$. It remains to show that the points p_i^ϵ are feasible for the inequalities (5.33b)–(5.33e), (5.35), (5.36), and (5.47). We consider four cases.

Case (a): $y + u < 0$. For $\epsilon > 0$, define $p_i^\epsilon = (x_i^\epsilon, u_i^\epsilon, y_i^\epsilon, z_i^\epsilon, t_i^\epsilon)$ where

$$x_i^\epsilon := (1 - \alpha_i^\epsilon) + \alpha_i^\epsilon x \qquad u_i^\epsilon := \overline{\gamma}(1 - \alpha_i^\epsilon) + \alpha_i^\epsilon u \qquad y_i^\epsilon := \alpha_i^\epsilon y$$
$$z_i^\epsilon := \alpha_i^\epsilon z \qquad t_i^\epsilon := (1 - \alpha_i^\epsilon)\overline{\gamma} + \alpha_i^\epsilon t$$

for $i = 1, 2$.

(5.33b) Satisfied strictly by p by the assumption of this case.

(5.33c)–(5.33e) Easily checked directly.

(5.35) Shown directly by

$$\overline{\gamma}x_i^\epsilon - u_i^\epsilon = \overline{\gamma}(1 - \alpha_i^\epsilon) + \overline{\gamma}\alpha_i^\epsilon x - (1 - \alpha_i^\epsilon)\overline{\gamma} - \alpha_i^\epsilon u = \alpha_i^\epsilon(\overline{\gamma}x - u) \geq 0. \quad (5.60)$$

(5.36) Shown directly by

$$u_i^\epsilon - \underline{\gamma}x_i^\epsilon = \overline{\gamma}(1 - \alpha_i^\epsilon) + \alpha_i^\epsilon u - \underline{\gamma}(1 - \alpha_i^\epsilon) - \underline{\gamma}\alpha_i^\epsilon x$$

$$= (\overline{\gamma} - \underline{\gamma})(1 - \alpha_i^\epsilon) + \alpha_i^\epsilon(u - \underline{\gamma}x) \quad (5.61)$$

$$\leq (\overline{\gamma} - \underline{\gamma})(1 - \alpha_i^\epsilon) + \alpha_i^\epsilon(t - \underline{\gamma})$$

$$= (\overline{\gamma} - \underline{\gamma})(1 - \alpha_i^\epsilon) + t_i^\epsilon - (1 - \alpha_i^\epsilon)\overline{\gamma} - \alpha_i^\epsilon\underline{\gamma} = t_i^\epsilon - \underline{\gamma}. \quad (5.62)$$

(5.47) Using (5.60) and (5.61), we get

$$(\overline{\gamma} - \underline{\gamma})y_i^\epsilon + \underline{\gamma}(\overline{\gamma}x_i^\epsilon - u_i^\epsilon) + \overline{\beta}(u_i^\epsilon - \underline{\gamma}x_i^\epsilon)$$

$$= \alpha_i^\epsilon\left((\overline{\gamma} - \underline{\gamma})y + \underline{\gamma}(\overline{\gamma}x - u) + \overline{\beta}(u - \underline{\gamma}x)\right) + \overline{\beta}(\overline{\gamma} - \underline{\gamma})(1 - \alpha_i^\epsilon)$$

$$\leq \alpha_i^\epsilon\overline{\beta}(t - \underline{\gamma}) + (1 - \alpha_i^\epsilon)\overline{\beta}(\overline{\gamma} - \underline{\gamma}) = \overline{\beta}(t_i^\epsilon - \underline{\gamma})$$

where the last equation follows from (5.62).

Case (b): $z + x < 1$. For $\epsilon > 0$, define $p_i^\epsilon = (x_i^\epsilon, u_i^\epsilon, y_i^\epsilon, z_i^\epsilon, t_i^\epsilon)$ where

$$x_i^\epsilon := \alpha_i^\epsilon x \qquad u_i^\epsilon := \alpha_i^\epsilon u \qquad\qquad y_i^\epsilon := \alpha_i^\epsilon y$$

$$z_i^\epsilon := \alpha_i^\epsilon z \qquad t_i^\epsilon := \alpha_i^\epsilon t + (1 - \alpha_i^\epsilon)\underline{\gamma}$$

for $i = 1, 2$.

(5.33b) Easily checked directly.

(5.33c) Satisfied strictly by p by the assumption of this case.

(5.33d), (5.33e), (5.35) Easily checked directly.

(5.36) Shown directly by

$$u_i^\epsilon - \underline{\gamma}x_i^\epsilon = \alpha_i^\epsilon u - \underline{\gamma}\alpha_i^\epsilon x \leq \alpha_i^\epsilon(t - \underline{\gamma}) = t_i^\epsilon - (1 - \alpha_i^\epsilon)\underline{\gamma} - \alpha_i^\epsilon\underline{\gamma}$$

$$= t_i^\epsilon - \underline{\gamma}. \quad (5.63)$$

(5.47) Shown directly by

$$(\overline{\gamma} - \underline{\gamma})y_i^\epsilon + \underline{\gamma}(\overline{\gamma}x_i^\epsilon - u_i^\epsilon) + \overline{\beta}(u_i^\epsilon - \underline{\gamma}x_i^\epsilon)$$

$$= \alpha_i^\epsilon\left((\overline{\gamma} - \underline{\gamma})y + \underline{\gamma}(\overline{\gamma}x - u) + \overline{\beta}(u - \underline{\gamma}x)\right)$$

$$\leq \alpha_i^\epsilon\overline{\beta}(t - \underline{\gamma}) = \overline{\beta}(t_i^\epsilon - \underline{\gamma})$$

where the last equation follows as in (5.63).

Case (c): $y - \bar{\beta}z < 0$ **and** $\underline{\beta}z - y < 0$. For $\epsilon > 0$, define $p_i^\epsilon = (x_i^\epsilon, u_i^\epsilon, y_i^\epsilon, z_i^\epsilon, t_i^\epsilon)$ where

$$x_i^\epsilon := \alpha_i^\epsilon x \qquad\qquad u_i^\epsilon := \alpha_i^\epsilon u \qquad\qquad y_i^\epsilon := \alpha_i^\epsilon y$$
$$z_i^\epsilon := (1 - \alpha_i^\epsilon) + \alpha_i^\epsilon z \qquad\qquad t_i^\epsilon := \alpha_i^\epsilon t + (1 - \alpha_i^\epsilon)\underline{\gamma}$$

for $i = 1, 2$. Then, it is easily seen by construction that p_i^ϵ satisfies (5.33c) for any $\epsilon \in (0, 1)$, $i = 1, 2$. As the definitions of x_i^ϵ, u_i^ϵ, y_i^ϵ, and t_i^ϵ are the same as in Case (b), the arguments of Case (b) apply for all inequalities that do not contain the variable z. This just leaves and (5.33d) and (5.33e), which by assumption are satisfied strictly by p, and so the proof for this case is complete.

Case (d): $y + u = 0$, $z + x = 1$, **and either** $y - \bar{\beta}z = 0$ **or** $\underline{\beta}z - y = 0$. For $\epsilon > 0$, define $p_i^\epsilon = (x_i^\epsilon, u_i^\epsilon, y_i^\epsilon, z_i^\epsilon, t_i^\epsilon)$ where

$$x_i^\epsilon := (1 - \alpha_i^\epsilon) + \alpha_i^\epsilon x \qquad\qquad u_i^\epsilon := \alpha_i^\epsilon u \qquad\qquad y_i^\epsilon := \alpha_i^\epsilon y$$
$$z_i^\epsilon := \alpha_i^\epsilon z \qquad\qquad t_i^\epsilon := (1 - \alpha_i^\epsilon)\frac{\underline{\gamma}\bar{\gamma}}{\bar{\beta}} + \alpha_i^\epsilon t$$

for $i = 1, 2$.

(5.33b)–(5.33e) Easily checked directly.

(5.35) We show that p satisfies (5.35) strictly. Suppose for purpose of contradiction that $\bar{\gamma}x - u = 0$. Then,

$$
\begin{aligned}
(\bar{\gamma} - \underline{\gamma})y + \underline{\gamma}(\bar{\gamma}x - u) + \bar{\beta}(u - \underline{\gamma}x) &= (\bar{\gamma} - \underline{\gamma})y + \bar{\beta}(\bar{\gamma}x - \underline{\gamma}x) \\
&= (\bar{\gamma} - \underline{\gamma})(y + \bar{\beta}x) = (\bar{\gamma} - \underline{\gamma})\bar{\beta} \\
&> \bar{\beta}(t - \underline{\gamma})
\end{aligned}
$$

where we have used $y + \bar{\beta}x = \bar{\beta}z + \bar{\beta}x = \bar{\beta}$. Thus, when $\bar{\gamma}x - u = 0$ then (5.47) is violated, and hence we conclude that (5.35) is satisfied strictly by p.

(5.36) We show that p satisfies (5.36) strictly. Indeed, as $y = -u$, we find that

$$(\bar{\gamma} - \underline{\gamma})y + \underline{\gamma}(\bar{\gamma}x - u) = (\bar{\gamma} - \underline{\gamma})(-u) + \underline{\gamma}(\bar{\gamma}x - u) = \bar{\gamma}(\underline{\gamma}x - u) > 0$$

since $\bar{\gamma} < 0$ and $\underline{\gamma}x - u < 0$. Thus, rearranging inequality (5.47) yields

$$u - \underline{\gamma}x \leq t - \underline{\gamma} - \frac{1}{\bar{\beta}}\left((\bar{\gamma} - \underline{\gamma})y + \underline{\gamma}(\bar{\gamma}x - u)\right) < t - \underline{\gamma}$$

which shows (5.36) is satisfied strictly by p.

(5.47) Shown directly by

$$
\begin{aligned}
&(\overline{\gamma} - \underline{\gamma})y_i^\epsilon + \underline{\gamma}(\overline{\gamma}x_i^\epsilon - u_\epsilon^i) + \overline{\beta}(u_i^\epsilon - \underline{\gamma}x_i^\epsilon) \\
&= \alpha_i^\epsilon\left((\overline{\gamma} - \underline{\gamma})y + \underline{\gamma}(\overline{\gamma}x - u) + \overline{\beta}(u - \underline{\gamma}x)\right) + \underline{\gamma}\overline{\gamma}(1 - \alpha_i^\epsilon) - \overline{\beta}\underline{\gamma}(1 - \alpha_i^\epsilon) \\
&\leq \alpha_i^\epsilon\overline{\beta}(t - \underline{\gamma}) - (1 - \alpha_i^\epsilon)\underline{\gamma}(\overline{\beta} - \overline{\gamma}) \\
&= \overline{\beta}\left(t_i^\epsilon - (1 - \alpha_i^\epsilon)\frac{\underline{\gamma}\overline{\gamma}}{\overline{\beta}} - \underline{\gamma}\alpha_i^\epsilon\right) - (1 - \alpha_i^\epsilon)\underline{\gamma}(\overline{\beta} - \overline{\gamma}) \\
&= \overline{\beta}(t_i^\epsilon - \underline{\gamma}) - (1 - \alpha_i^\epsilon)(\underline{\gamma}\overline{\gamma}) + (1 - \alpha_i^\epsilon)(\underline{\gamma}\overline{\gamma}) = \overline{\beta}(t_i^\epsilon - \underline{\gamma}).
\end{aligned}
$$

$\square$

Proof of Theorem 5.16

We now state and prove the two main lemmas that support the proof of Theorem 5.16. We prepare the proofs with the following proposition.

Proposition 5.25: *Let $\underline{\beta} < 0$. If $p \in R^3$, then p satisfies the following inequality:*

$$
(\underline{\gamma} - \underline{\beta})x \leq -\underline{\beta} \tag{5.64}
$$

In addition, if p satisfies (5.33b), (5.33c), (5.33e), and (5.34) at equality, then it satisfies (5.64) at equality.

Proof. First, aggregating (5.33b) with weight 1, (5.33c) with weight $-\underline{\beta}$, (5.33e) with weight 1, and (5.34) with weight -1, yields (5.64). If (5.33b), (5.33c), (5.33e), and (5.34) are all satisfied at equality, then p satisfies (5.64) at equality. $\square$

Lemma 5.26: *Suppose $0 < \underline{\gamma} < \overline{\gamma}$ and $\underline{\beta} < 0 < \overline{\beta}$. Let $p = (x, u, y, z, t) \in R^3$ with $0 < x < 1$ and $\underline{\gamma} < t < \overline{\gamma}$. If $u < xt$, then p is not an extreme point of R^3.*

Proof. This proof has the same structure as the proof of Lemma 5.21. By Proposition 5.19, p satisfies (5.35), (5.36), and (5.40) with strict inequality. Also, as $u \geq \underline{\gamma}x > 0$, (5.33b) implies that $y < 0 \leq \overline{\beta}z$, and hence p satisfies (5.33d) with strict inequality. In addition, by (5.53),

$$
\overline{\gamma}x - u \leq \frac{-\underline{\beta}}{\underline{\gamma} - \underline{\beta}}(\overline{\gamma} - t) < \overline{\gamma} - t
$$

as $\overline{\gamma} - t > 0$ and $\underline{\gamma} - \underline{\beta} > -\underline{\beta}$ because $\underline{\gamma} > 0$, and so p satisfies (5.37) with strict inequality. It remains to show that the points p_i^ϵ satisfy (5.33b), (5.33c), (5.33e), (5.34), and (5.53) for ϵ small enough. We consider four cases.

Case (a): $y + u < 0$. For $\epsilon > 0$, define $p_i^\epsilon = (x_i^\epsilon, u_i^\epsilon, y_i^\epsilon, z_i^\epsilon, t_i^\epsilon)$ where

$$x_i^\epsilon := \alpha_i^\epsilon x \qquad u_i^\epsilon := \alpha_i^\epsilon u \qquad y_i^\epsilon := (1 - \alpha_i^\epsilon)\underline{\beta} + \alpha_i^\epsilon y$$

$$z_i^\epsilon := (1 - \alpha_i^\epsilon) + \alpha_i^\epsilon z \qquad t_i^\epsilon := (1 - \alpha_i^\epsilon)\overline{\gamma} + \alpha_i^\epsilon t$$

for $i = 1, 2$.

(5.33b) Satisfied strictly by p by the assumption of this case.

(5.33c), **(5.33e)**, **(5.34)** Easily checked directly.

(5.53) Shown directly by

$$(\underline{\gamma} - \underline{\beta})(\overline{\gamma} x_i^\epsilon - u_i^\epsilon) = \alpha_i^\epsilon (\underline{\gamma} - \underline{\beta})(\overline{\gamma} x - u)$$
$$\leq \alpha_i^\epsilon (-\underline{\beta})(\overline{\gamma} - t)$$
$$= -\underline{\beta}(\alpha_i^\epsilon \overline{\gamma} - t_i^\epsilon + (1 - \alpha_i^\epsilon)\overline{\gamma}) = -\underline{\beta}(\overline{\gamma} - t_i^\epsilon).$$

Case (b): $z + x < 1$. For $\epsilon > 0$, define $p_i^\epsilon = (x_i^\epsilon, u_i^\epsilon, y_i^\epsilon, z_i^\epsilon, t_i^\epsilon)$ where

$$x_i^\epsilon := \alpha_i^\epsilon x \qquad u_i^\epsilon := \alpha_i^\epsilon u \qquad y_i^\epsilon := \alpha_i^\epsilon y$$

$$z_i^\epsilon := \alpha_i^\epsilon z \qquad t_i^\epsilon := (1 - \alpha_i^\epsilon)\overline{\gamma} + \alpha_i^\epsilon t$$

for $i = 1, 2$. Then, p_i^ϵ is easily seen to satisfy (5.33b), (5.33e), and (5.34) for any $\epsilon \in (0, 1)$. (5.33c) is satisfied strictly by the assumption of this case. In addition, as the definitions of u_i^ϵ, s_i^ϵ, and t_i^ϵ are the same as in Case (a), (5.53) is satisfied by p_i^ϵ for $i = 1, 2$ and any $\epsilon \in (0, 1)$.

Case (c): $y > \underline{\beta} z$. For $\epsilon > 0$, define $p_i^\epsilon = (x_i^\epsilon, u_i^\epsilon, y_i^\epsilon, z_i^\epsilon, t_i^\epsilon)$ where

$$x_i^\epsilon := \alpha_i^\epsilon x \qquad u_i^\epsilon := \alpha_i^\epsilon u \qquad y_i^\epsilon := \alpha_i^\epsilon y$$

$$z_i^\epsilon := (1 - \alpha_i^\epsilon) + \alpha_i^\epsilon z \qquad t_i^\epsilon := (1 - \alpha_i^\epsilon)\overline{\gamma} + \alpha_i^\epsilon t$$

for $i = 1, 2$. Then, p_i^ϵ is easily seen to satisfy (5.33b), (5.33c), and (5.34) for any $\epsilon \in (0, 1)$. (5.33e) is satisfied strictly by the assumption of this case. In addition, as the definitions of u_i^ϵ, x_i^ϵ, and t_i^ϵ are the same as in Case (a), (5.53) is satisfied by p_i^ϵ for $i = 1, 2$ and any $\epsilon \in (0, 1)$.

Case (d): $y + u = 0$, $z + x = 1$, **and** $y = \underline{\beta} z$. For $\epsilon > 0$, define $p_i^\epsilon = (x_i^\epsilon, u_i^\epsilon, y_i^\epsilon, z_i^\epsilon, t_i^\epsilon)$ where

$$x_i^\epsilon := x - \delta_i^\epsilon \qquad u_i^\epsilon := u - \underline{\beta}\delta_i \qquad y_i^\epsilon := y + \underline{\beta}\delta_i$$

$$z_i^\epsilon := z + \delta_i \qquad t_i^\epsilon := t + \delta_i \frac{(\underline{\gamma} - \underline{\beta})(\overline{\gamma} - \underline{\beta})}{(-\underline{\beta})}$$

for $i = 1, 2$.

(5.33b), **(5.33c)**, **(5.33e)** Easily checked directly.

(5.34) We show that (5.34) is satisfied strictly by p. Indeed, suppose to the contrary that $\underline{\gamma} x - u = 0$. Then, by Proposition 5.25, $(\underline{\gamma} - \underline{\beta}) x = -\underline{\beta}$. Thus, using $u < xt$, $\underline{\gamma} > 0$ and $-\underline{\beta} > 0$,

$$(\underline{\gamma} - \underline{\beta})(\overline{\gamma} x - u) > (\underline{\gamma} - \underline{\beta})(\overline{\gamma} x - xt) = -\underline{\beta}(\overline{\gamma} - t)$$

and hence (5.53) is violated. Thus, (5.34) is satisfied strictly by p.

(5.53) Since $u > xt$ and Proposition 5.25 we show the validity of (5.53) by

$$
\begin{aligned}
(\underline{\gamma} - \underline{\beta})(\overline{\gamma} x_i^{\epsilon} - u_i^{\epsilon}) &= (\underline{\gamma} - \underline{\beta})\big(\overline{\gamma}(x - \delta_i) - (u - \underline{\beta}\delta_i)\big) \\
&= (\underline{\gamma} - \underline{\beta})(\overline{\gamma} x - u) - \delta_i(\underline{\gamma} - \underline{\beta})(\overline{\gamma} - \underline{\beta}) \\
&< (\underline{\gamma} - \underline{\beta}) x(\overline{\gamma} - t) - \delta_i(\underline{\gamma} - \underline{\beta})(\overline{\gamma} - \underline{\beta}) \\
&\le -\underline{\beta}(\overline{\gamma} - t) - \delta_i(\underline{\gamma} - \underline{\beta})(\overline{\gamma} - \underline{\beta}) \\
&= -\underline{\beta}\left(\overline{\gamma} - t_i + \delta_i \frac{(\underline{\gamma} - \underline{\beta})(\overline{\gamma} - \underline{\beta})}{(-\underline{\beta})}\right) - \delta_i(\underline{\gamma} - \underline{\beta})(\overline{\gamma} - \underline{\beta}) \\
&= -\underline{\beta}(\overline{\gamma} - t_i).
\end{aligned}
$$

$\square$

Lemma 5.27: *Suppose $0 < \underline{\gamma} < \overline{\gamma}$ and $\underline{\beta} < 0 < \overline{\beta}$. Let $p = (x, u, y, z, t) \in R^3$ with $0 < x < 1$ and $\underline{\gamma} < t < \overline{\gamma}$. If $u > xt$, then p is not an extreme point of R^3.*

Proof. Using $\underline{\gamma} - \underline{\beta} > 0$, we have

$$(\underline{\gamma} - \underline{\beta})(x\overline{\gamma} - u) < (\underline{\gamma} - \underline{\beta}) x(\overline{\gamma} - t) \le -\underline{\beta}(\overline{\gamma} - t)$$

by (5.64) in Proposition 5.25.

When $u > xt$, the inequality (5.53) is satisfied with strict inequality, just as (5.43) is satisfied by strict inequality when $u > xt$ and $\underline{\gamma} < 0$ as in Case 1. As the substitution of (5.53) for (5.43) is the only difference between the sets R^3 and R^1, Lemma 5.22 applies directly to this case, and we can conclude that if $u > xt$, $0 < x < 1$, and $\underline{\gamma} < t < \overline{\gamma}$, then p is not an extreme point of R^3. $\square$

This completes the proofs of the convex hull for the cases we considered.

5.4 Computational experiments

In this section, we show that the presented inequalities indeed strengthen the McCormick relaxation of the *pq*-formulation on a large testset of pooling instances. To this end, we conduct computational experiments on instances from the literature and on randomly generated instances. We show that proposed inequalities indeed strengthen the relaxation of the pq-formulation and are able to speed up the global solution process, especially on sparse instances.

5.4.1 Computational setup

The experiments were conducted on a cluster with 64bit Intel Xeon X5672 CPUs at 3.2 GHz with 12 MB cache and 48 MB main memory. To avoid random noise, e.g., by cache misses, we run only one job on each node at a time.

The model is implemented in the GAMS language [GAMa] and processed with GAMS version 24.7.1. The pq-formulation is solved to global optimality with SCIP version 3.2 which used CPLEX 12.6.3 as LP solver and Ipopt 3.12 as local NLP solver. The relaxations, which are LPs or SOCPs, are solved with CPLEX 12.6.3. We used the predefined timelimit of 1000 seconds and use a relative gap of 10^{-6} as termination criterion (GAMS options OPTCA $= 0.0$ and OPTCR $= 10^{-6}$).

5.4.2 Adding the inequalities

Recall that the initial step to construct the 5-variable relaxation was to focus on a (Spec, Pool, Output) pair and extend the model by the aggregated variables u, x, z, y, t for each such pair. This is the road we also take in the implementation. We extend the model by the aggregated variables and rely on the solver to replace to disaggregate the variables in the constraints if it is considered advantageous. We add the linear inequalities whenever they are valid (specifically (5.47) is added whenever $\overline{\beta} > 0$ and (5.53) is added whenever $\underline{\beta} < 0$). Inequality (5.40) is second-order cone representable and could in principle be added directly as SOC constraint. This becomes problematic when (5.43) comes into play. There is no linear or second-order cone representation known and we thus resort to a cutting plane algorithm. Namely, whenever the relaxation solution has $y > 0$ for a specific (Spec, Pool, Output), a gradient inequality at this point is separated (note that the gradient inequality is also valid for $y \leq 0$ due to the convex extension that is valid for T) and the relaxation is solved again. Since the gradient inequalities towards the end of the separation loop become almost parallel, the interior point SOCP solver frequently runs into numerical trouble. To circumvent this, (5.40) is not added directly to the model, but linear gradient inequalities are also separated from all conic inequalities in the same separation loop. The major advantage is that all relaxations are then LPs and thus solved very efficiently. This approach in our experience provides much better running times than solving SOCP relaxations in the separation loop and we did not experience numerical

Inputs I Pools L Outputs J

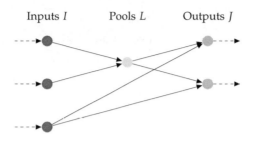

Figure 5.4: The original network from Haverly.

inconsistencies.

We separate the inequalities only at the root node of the spatial branch-and-bound algorithm. More precisely, we setup the separation loop for both inequalities and separate until the absolute violation of the conic inequality and inequality (5.43) are below 10^4 and 10^5, respectively. Then we pass the pq-formulation and all inequalities that have been separated to SCIP and solve the problem globally. The separation therefore does take any model changes or strengthening from the preprocessing or from propagations during the cutting plane loop in SCIP into account.

In the following we will refer with pq-relaxation to the McCormick relaxation of the pq-formulation. The relaxation that arises by strengthening the pq-relaxation with out valid inequalities is called pq^+-relaxation.

5.4.3 Instances

We perform experiments on two sets of instances: The pooling instances from the GAMSLIB [GAMb] and randomly generated instances. The GAMSLIB instances are encoded in the `pool` model as different cases totaling up to 14 instances. All of them were first presented in scientific publications about the pooling problem. It comprises 3 instances on the original network from Haverly [Hav78; Hav79] which is depicted in Fig. 5.4. Furthermore, it contains instances from the publications [FHJ92; BEG94; ATS99; Aud+04].

The random instances are generated in the following way. The basis are copies of the Haverly instances. The resulting disconnected graphs are then supplemented by randomly adding a specific number of admissible edges in a pooling network. As the resulting network might still be disconnected, the first edges are chosen as to connect two disconnected components until the graph is

Instance	Graph		pq		pq^+		Opt
	Nodes	Arcs	Absolute	Gap	Absolute	Gap	
adhya1	11	13	-766.3	39.4 %	-697.0	26.8 %	-549.8
adhya2	11	13	-570.8	3.8 %	-568.3	3.4 %	-549.8
adhya3	15	20	-571.3	1.8 %	-570.7	1.7 %	-561.0
adhya4	15	18	-961.2	9.5 %	-955.4	8.9 %	-877.6
bental4	7	4	-550.0	22.2 %	-450.0	0.0 %	-450.0
haverly1	6	6	-500.0	25.0 %	-400.0	0.0 %	-400.0
haverly2	6	6	-1000.0	66.7 %	-600.0	0.0 %	-600.0
haverly3	6	6	-800.0	6.7 %	-791.7	5.6 %	-750.0

Table 5.3: Results on GAMSLIB instances where the pq-relaxation does not provide the optimum

connected.

The GAMSLIB includes 3 instances on the Haverly network with different parameters and the distribution among the 3 Haverly instances is sampled randomly. Next, for each copy a factor $\phi \in [0.5, 2]$ is sampled uniformly and all concentration parameters, i.e., λ_{ik} and $\overline{\mu}_{jk}$ of that copy are scaled by ϕ. Lower bounds on the concentration are not used in these instances but could be sampled and handled in the separation in a similar way.

We generated instances with 10, 15, and 20 copies of the Haverly network. The number of edges to be added are multiples of number of copies of the Haverly network. For each such pair of number of copies and number of additional edges, we sample 10 instances. In total 180 instances are generated.

The Haverly instances have only one attribute. To test the sensitivity of the proposed relaxation w.r.t. the number of attributes, we generate additional instances based on the instances above where we add an additional randomly generated attribute. To add the attribute, we first assign concentrations to inputs and assume a uniform distribution in $[0, 10]$. Then we sample upper bounds on the concentration of the attribute for each output from a uniform distribution on the interval of the minimum and maximum concentrations of the reachable inputs. Clearly, this doubles the number of instances to be solved to 360.

5.4.4 Results

First, we consider the 14 GAMSLIB instances. For 6 instances the pq-relaxation provides the optimal bound and hence these instances are not considered anymore. Table 5.3 shows results on the remaining 8 GAMSLIB instances where the pq-relaxation does not already provide the optimum value. Along with the size of the graph in terms of number of nodes and arcs, Table 5.3 contains the value of the different relaxations and their gaps. The column "Opt" shows the global optimum computed by solving the non-convex pq-formulation. The instances are relatively small, but the results are encouraging. Our relaxation gives a stronger dual bound on all instances compared to the pq-relaxation and on 3 instances the gap is closed completely.

Table 5.4 presents results on the randomly generated networks. The table is split into two parts: One for the instances with 1 attribute and a second with the instances with 2 attributes. Both blocks are terminated by average values of all instances in the block and the last line in the table is the Grand Total over all instances. The instances are grouped by the number of copies of the Haverly network (first column) and by the number of edges that have been added to the network (column $|A^+|$). Each row thus provides aggregated results over 10 instances. The group of columns labeled with "Graph" shows statistics about the graphs. Besides the number of random arcs added $|A^+|$, the number of nodes $|V|$ and arcs $|A|$ is shown. The numbers are identical with each group of instances. Next, the average gap for the pq-relaxation and the pq^+-relaxation is shown. For both approaches the gap is computed w.r.t. the best known primal bound for the problem and thus reflects only differences in the dual bound. Finally, the last two groups of columns show statistics about the global solution process using the pq-formulation and using the strengthened pq^+-relaxation at the root. We report the number of instances that ran into the timelimit (column "TL"), time and number of nodes. For time and nodes, the shifted geometric mean with shift 2 and 100, respectively, is used to aggregate the results. Furthermore, only instances where both approaches finished within the timelimit are considered in the computation of the number of nodes. For pq^+, the separation time for the nonlinear inequalities is taken into account by adding it to the time SCIP needed to solve the problem. The appendix of the online version of this thesis [Sch17] provides detailed performance figures for all instances.

The pq^+-relaxation is very effective in reducing the root gap leaving a gap of 3.0 % compared to the 5.5 % of the pq-relaxation over all instances in the shifted geometric mean. Especially for small densities the pq^+-relaxation performs extremely well. This is expected, since the relaxation provides the optimal dual bound on two out of the three Haverly instances (see Table 5.3) that we used to construct the random instances. All but one instance of the testset experience an improvement of the dual bound due to the additional inequalities. When looking at the global optimization, the most notable effect of the stronger root bound is

Cop.	Graph			Gap [%]		Global pq			Global pq^+		
	$|V|$	$|A|$	$|A^+|$	pq	pq^+	TL	Time	Nodes	TL	Time	Nodes
Instances with 1 attribute											
10	60	70	10	13.0	3.2	0	5.0	7098.1	0	1.8	213.9
	60	80	20	8.3	3.8	0	3.4	3011.6	0	3.4	727.0
	60	90	30	4.7	2.9	0	3.1	1397.2	0	3.2	371.4
	60	100	40	3.0	1.9	0	2.3	993.2	0	4.1	530.4
	60	110	50	2.6	2.1	0	3.3	1034.4	0	6.0	665.9
	60	120	60	3.3	2.4	0	6.3	2320.4	0	9.3	1511.6
15	90	105	15	10.6	3.2	0	63.0	106880.6	0	7.0	2023.2
	90	120	30	7.2	3.3	1	53.2	40031.5	1	20.0	3480.5
	90	135	45	4.9	3.4	1	36.6	24087.1	0	31.0	8202.0
	90	150	60	4.1	3.0	1	33.4	19234.3	0	24.0	5043.2
	90	165	75	3.3	2.5	0	21.2	13919.3	0	37.1	10576.1
	90	180	90	3.7	3.0	1	47.4	21998.2	1	51.8	8300.6
20	120	140	20	13.4	4.3	9	993.9	1655439.0	1	44.3	7327.0
	120	160	40	6.0	2.9	4	296.9	175642.9	3	116.8	18319.4
	120	180	60	4.5	2.8	5	287.6	84123.7	4	213.3	29497.7
	120	200	80	4.1	2.8	2	84.3	40476.0	2	68.5	12186.5
	120	220	100	3.1	2.3	3	159.1	44945.7	3	142.3	20325.7
	120	240	120	2.5	2.0	2	187.5	69610.8	3	224.1	36090.3
Total	–	–	–	5.7	2.9	29	37.1	12685.7	18	25.0	3108.0
Instances with 2 attributes											
10	60	70	10	9.2	3.2	0	1.8	843.0	0	1.9	147.2
	60	80	20	6.3	3.7	0	2.0	753.4	0	2.9	274.1
	60	90	30	5.1	3.7	0	1.8	930.2	0	2.9	381.2
	60	100	40	4.0	2.6	0	4.3	2010.4	0	10.9	1992.5
	60	110	50	3.5	2.8	0	8.1	3299.5	0	11.8	2376.6
	60	120	60	4.7	3.6	1	8.1	2275.9	1	13.7	1478.4
15	90	105	15	6.8	2.6	0	3.8	2554.7	0	5.6	275.2
	90	120	30	6.3	3.1	1	7.8	1927.5	1	9.7	411.2
	90	135	45	5.0	3.6	1	16.7	5807.5	0	19.8	2776.0
	90	150	60	4.8	3.5	1	42.9	26362.6	0	43.3	10231.2
	90	165	75	5.8	4.6	0	39.7	30319.8	0	66.9	18086.5
	90	180	90	4.4	3.5	3	67.1	10681.1	2	104.9	6737.0
20	120	140	20	8.4	3.0	0	22.4	19759.7	0	10.3	1223.9
	120	160	40	4.6	2.4	0	19.1	17450.4	0	15.5	3286.8
	120	180	60	5.5	3.8	2	58.9	15155.9	2	85.4	7711.4
	120	200	80	4.3	3.1	3	49.1	7361.7	3	75.5	4646.3
	120	220	100	3.6	2.6	0	22.2	7413.0	1	38.2	3819.4
	120	240	120	3.3	2.8	2	75.8	12039.9	3	92.6	5384.1
Total	–	–	–	5.3	3.2	14	15.1	5026.2	13	19.5	1893.5
Grand Total	–	–	–	5.5	3.0	43	23.9	7832.5	31	22.1	2402.2

Table 5.4: Results on randomly generated instances.

on the number of branch-and-bound nodes needed to solve an instances to global optimality. The shifted geometric mean of the nodes is reduced from 7832 to 2402, a reduction of 69 % on all instances. While reductions are stronger on sparse instances, significant reductions are observed among all classes of instances. In terms of time to optimality, the shifted geometric mean over all instances decreases only slightly from 23.9 to 22.1 seconds by using the pq^+-relaxation, but big differences exist between the instances with one and two attributes.

On the instances with one attribute, the shifted geometric mean of the runtime decreases from 37.1 to 25.0 seconds and the number of timeouts is reduces from 29 to 18 by using the pq^+-relaxation. The biggest winners are sparse instances and as the instances become denser the pq-formulation achieves better running times in the shifted geometric mean. A large portion of the overall improvement on the instances with 1 attribute comes from instances with 20 Haverly networks and only 20 additional edges. From the 10 instances of this class, only one instance is solved within the timelimit (in 940 seconds) by the pq-formulation while all but one are solved using the pq^+-relaxation. The dual bound is exactly the problem on these instances. The approach with the pq-formulation found an optimal solution always within the first 184 seconds of the optimization and then used a massive amount of branching nodes to close the gap.

On the instances with two attributes, the shifted geometric mean of the runtime increases from 15.1 to 19.5 seconds by using the pq^+-relaxation. In general, the pq-relaxation benefits much more from the additional attribute than the pq^+-relaxation does. Especially on sparse instances, the additional constraints introduced by the presence of a second attribute make the pq-relaxation much tighter. For example on instances with 20 copies of the Haverly network and 20 additional edges, the gap is reduced from 13.4 % to 8.4 % with the additional attribute and now all instances in this group are solved in 22.4 seconds in the shifted geometric mean. As the pq^+-relaxation strengthens the pq-relaxation, the pq^+-relaxation also benefits from a tighter pq-relaxation, but our inequalities are not as effective with the additional attribute as the gap increases from 2.9 % to 3.2 % for the pq^+-relaxation in the shifted geometric mean.

Figure 5.5 shows scatter plots of the running times for instances with 1 and 2 attributes. The time to global optimality using the pq-formulation is plotted on the x-axis and the time using the strengthened pq^+-relaxation on the y-axis such that points below the diagonal are favorable for the pq^+-relaxation. Different markers are used to differentiate the instances with different numbers of Haverly networks. Circles are used for instances with 10 Haverly networks, diamonds for 15 and squares for 20 networks. The color represents the density of the instances and warmer colors represent higher density. The density is computed w.r.t. all admissible pooling arcs. Of course, graphs with more copies of the Haverly network tend to have lower densities as the number of admissible edges increases in the order of the number of copies already present while the number of edges in an additional copy is constant. For 1 attribute, the plot in Fig. 5.5a confirms that on graphs with higher density, the pq-formulation has advantages, where on sparse graphs using the pq^+-relaxation gives some neat speed-ups. For 2 attributes, the plot in Fig. 5.5b also shows that the pq^+-relaxation tends to perform better on sparse instances, but the speed-ups are not as high, especially since the pq-formulation performs much better.

5.5 Conclusion

The pooling problem is a very challenging problem with application in several domains. The underlying network structure as well as the nonconvexity introduced by material blending provide a rich structure than can be exploited. In this chapter, after reviewing the standard formulations and relaxations of the problem, we use the state-of-the-art pq-formulation as starting point to derive a custom relaxation. This is done by reducing the network to its minimal essence comprising only an aggregated version of the inputs, one output, one pool and one attribute. At the same time, the relaxation still contains one bilinear term and captures the central non-convex structure of the model. The small size of this relaxation allowed us to completely describe its convex hull which in general is not a polyhedron. We show a complete inner description in terms of the extreme points and a complete outer description in terms of valid convex inequalities. The valid inequalities are added directly or by means of the separation of gradient inequalities to the pq-formulation.

The question whether the proposed relaxation is tight enough to improve the McCormick relaxation of the pq-formulation was answered positively in extensive computational experiments. The additional inequalities improve the pq-relaxation on all pooling instances from the GAMSLIB where the pq-formulation does not already provide the optimum value and provides the value of the optimal solution on 3 instances. On a large testset of randomly generated instances, the proposed inequalities reduce the shifted geometric mean of the gap between the relaxation value and the best known feasible solution from 5.5 % for the pq-relaxation to 3.0 % with the additional inequalities. Across all instances, the number of branch-and-bound nodes needed for a global solution is reduced considerably. In terms of time to optimality, the proposed relaxation performs very well on sparse instances, especially many more instances with only one attribute were solved to optimality within the timelimit.

Overall, we conclude that the formulation of a relaxation that maintains some of the structural properties of the original problem is an important step in this work. The study of the structure of the relaxation then enabled us to provide new valid inequalities for the pooling problem that have an impact on practical computations.

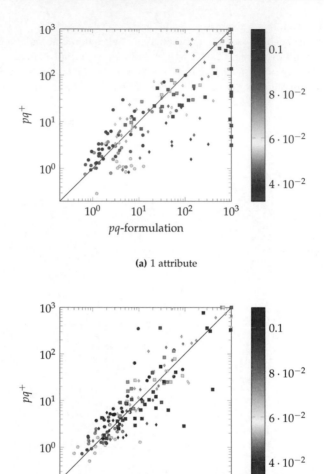

(a) 1 attribute

(b) 2 attributes

Figure 5.5: Scatter plot that compares the time to global optimality for the pq-formulation and using the pq^+-relaxation. Markers represent the number copies of the Haverly network (circle: 10, diamond: 15, square: 20). Colors represent the density of the graph.

6 Models for deterministic gas network optimization

Recent changes in the regulation of the German gas market are creating new challenges for *Transmission System Operators* (*TSO*). Especially the unbundling of gas transport and trading reduces the influence of network operators on transportation requests. Greater flexibility in the network is therefore demanded and the networks have to be extended accordingly. In this chapter, we describe the mathematical optimization model for topology planning in a deterministic setting.

The models and methods presented in this chapter have been developed within the FORNE project[1]. FORNE was a joint research project of Europe's largest gas network operation Open Grid Europe GmbH and several academic institutions. Credits for the work in this chapter go to colleagues from all partners in the FORNE project. We present it here in preparation for our approach for network planning considering a set of scenarios.

We restrict our presentation to a level needed to understand the mathematical structure of the problem. The reader is referred to [Koc+15; Pfe+14] for further details on the assumptions underlying our model and for precise formulas for the coefficients and [Füg+11; Hum14] for on overview over deterministic gas network extension.

This chapter is organized as follows. Section 6.1 introduces and motivates the problem. In Section 6.2 we sketch the MINLP model of a gas network before turning our attention to the network extension problem in Section 6.3.

6.1 Introduction

Gas transmission networks are complex structures that consist of passive elements and active, controllable elements such as valves and compressors. The behavior of the passive elements is entirely governed by the laws of physics and the network operator has no means to influence that behavior. Pipes are

---

[1]http://www.zib.de/projects/forne-research-cooperation-network-optimization

the most important representative of that group. Other passive elements are for example measurement equipment that causes a pressure drop or artificial resistors modeling e.g., extensive piping within stations. Active elements on the other hand are controlled by the network operator. Several active elements exist: Valves can be open or close and are a means to decouple different parts of the network. Compressors and control valves allow to increase and decrease the pressure within technical limitations. For planning purposes, the relationship of flow through a pipe and the resulting pressure difference is appropriately modeled by a nonlinear equation. The description of the active elements on the other side involves discrete decisions, e.g., whether a valve is open or closed. Therefore, the model to describe a gas network is a MINLP and its feasible set is non-convex in general.

Determining best possible network extensions at minimum cost is a difficult task as the network can be extended in various ways. In principle any two points can be connected by a new pipe and pipes are available in different standardized diameters. Building an additional pipe next to an existing one is called *looping*. Loops are the favorite extensions of network operators as they are considerably cheaper than new pipes as the regulatory process is much simpler and the TSO most often already owns the land the pipe is built on. In addition to pipes, new active elements can also be added anywhere in the network. Hence, generating meaningful extension candidates is a challenging task on its own. We assume extension candidates already given as part of the input to the problem.

While typically network extensions are increasing the transport capacity of a network, they can also cause new infeasibilities. A new pipe allows flow but couples the pressures at the end nodes, possibly rendering previously feasible transport requests infeasible. An additional valve retains all possibilities of the original network. Closing the valve forbids flow over the pipe which effectively removes the pipe from the system. The valve as a results ensures that the extension only adds features to the network. The same holds for compressors and control valve that in addition to bypass and closed states also have an active state. The bypass and closed state behave the same as an open or closed valve. So these active elements also supersede valves in features. Overall the extensions for a hierarchy where more complex and expensive extensions have all features of smaller and less expensive ones. This structure of the extensions will be exploited in the design of our decomposition approach in the next chapter.

6.2 Modeling gas transportation networks

The gas network is modeled as a directed graph $G = (V, A)$, where the arcs A are physical network components. Within the network, gas is to be transported from entries to exits. The flow at these points if given by a so-called *nomination* and is modeled by a vector $q^{\text{nom}} \in \mathbb{R}^V$ where positive and negative values of q_u^{nom} means that flow is leaving and entering the network at node u, respectively.

We assume a steady-state model where dynamic effects are not taken into account and pressure within the arcs is assumed to be constant. Therefore, we introduce pressure variables p_u to track the pressure at node $u \in V$. Flow though an element is modeled by a variable q_a for arc $a \in A$. Flows in the direction of the arcs are encoded by positive values for q_a while negative values encode flow in the reverse direction. We assume a homogeneous gas composition. Under this assumption, gas blending effects at the nodes can be ignored and the flow respects the flow conservation constraints

$$\sum_{a \in \delta^+(u)} q_a - \sum_{a \in \delta^-(u)} q_a = q_u^{\text{nom}},$$

where $\delta^-(u)$ and $\delta^+(u)$ are the arcs entering and leaving node u, respectively. Flow conservation constraints at all nodes ensure that the flow is compliant with the given nomination. Additionally, flow and pressure can have technical upper and lower bounds.

The relationship between flow and pressure depends on the network element, i.e., on the type of the arc. Generally, network elements can be partitioned into two groups: Passive and active elements. In the following, we will shortly review the model for pipes as the most prominent representative of a passive element and the models of the different active elements.

Pipes. Pipes are used to transport gas over long distances. A difference in the pressures in the end points is responsible for gas to flow. Mathematically, the relationship between the pressure at the end nodes u and v of a pipe a and the flow q_a is modeled by the equation

$$\alpha_a |q_a| q_a = p_u^2 - \beta_a p_v^2 \tag{6.1}$$

The parameters α_a and β_a are determined by the properties of the pipe and are assumed to be constant.

The right hand side can be linearized by reformulating it using variables for the square of the pressure $\pi_u = p_u^2$. Since the pressure is always positive (above the atmospheric pressure), this reformulation does not introduce ambiguities.

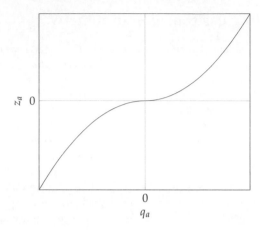

Figure 6.1: Nonlinear equation $|q_a|q_a = z_a$

Furthermore, we introduce an auxiliary variable z_a and split the equation into a linear and a nonlinear equation:

$$\alpha_a |q_a| q_a = z_a \tag{6.2}$$
$$z_a = \pi_u^2 - \beta_a \pi_v^2. \tag{6.3}$$

The only remaining nonlinearity is then present in (6.2), see Fig. 6.1 for a plot. If the pressure variable p_u is needed to model a network element, it is added to the model together with the coupling constraint $\pi_u = p_u^2$, otherwise it is omitted.

Active Elements: Valve, control valve, and compressor. The three most important active elements are valves, control valves and compressors. Valves can be used to disconnect parts of the network. Control valves have the additional feature that they are able to reduce the pressure while compressors can increase the pressure. In contrast to pipes, whose behavior is completely ruled by gas physics, the state of active elements can be controlled by the network operator.

The simplest active element is a valve. Valves have two possible states, open and closed, which is modeled by a binary variable s_a. An open valve ($s_a = 1$) does not cause a change in the pressure and allows flow within some technical bounds. A closed valve ($s_a = 0$) does not allow flow, but completely decouples the pressures at both end points. Mathematically, the conditions for a valve

$a = (u, v)$ can be expressed as follows:

$$s_a = 0 \quad \Rightarrow \quad q_a = 0 \tag{6.4}$$

$$s_a = 1 \quad \Rightarrow \quad p_u = p_v \tag{6.5}$$

These implications can be implemented using indicator or so-called Big-M constraints.

In addition to open (called "bypass") and closed, control valves and compressors have an additional possible state: The active state. When in this state, the actual increase or decrease of pressure takes place. Compressors and control valves use two binary variables s_a^{bp} and s_a^{ac} to decide the state of the element. If $s_a^{bp} = 1$, then the element is in bypass mode. If $s_a^{ac} = 1$, the element is in active mode. If both variables are zero, the element is closed.

A control valve $a = (u, v)$ allows the reduction of the pressure in direction of the flow within certain bounds when in active state. The constraints for a control valve are thus:

$$s_a^{bp} = 0, \, s_a^{ac} = 0 \quad \Rightarrow \quad q_a = 0 \tag{6.6}$$

$$s_a^{bp} = 1, \, s_a^{ac} = 0 \quad \Rightarrow \quad p_u = p_v \tag{6.7}$$

$$s_a^{bp} = 0, \, s_a^{ac} = 1 \quad \Rightarrow \quad \begin{cases} q_a \geq 0 \\ 0 < \underline{\Delta}_a \leq p_u - p_v \leq \overline{\Delta}_a \end{cases} \tag{6.8}$$

$$s_a^{bp} + s_a^{ac} \leq 1 \tag{6.9}$$

The equations (6.6)–(6.8) describe the three states. Inequality (6.9) makes sure that exactly one of the three states is selected.

Compressors are by far the most complex elements. The pressure increase depends on the flow and is governed by the so-called *characteristic diagram* which typically is a non-convex set. We use a linear approximation to the characteristic diagram and remain with the statement that the triple (p_u, p_v, q_a) must be in a certain polytope.

Since bypass and closed state of compressors and control valves behave identically to a valve, we also model valves with the two binary variables s_a^{bp} and s_a^{ac}, but fix s_a^{ac} to zero.

The question whether a network allows feasible operation for a given nomination q^{nom} is called *nomination validation*. Nomination validation is a challenging task on its own which network operators routinely face in daily operation as well as tactical and strategic planning. Formulated in this way, it is a feasibility problem without objective function. We refer to [Koc+15; Pfe+14] for a more detailed description of the network elements and their coefficients as well for details on the nomination validation problem.

6.3 Deterministic network extension

In this section, we extend the feasibility problem of checking whether a nomination allows a feasible operation in a network to the selection of a cost-optimal set of network extensions that allows the operation of a previously infeasible nomination. More details on the approach for deterministic network extension can be found in [Füg+11; Hum14].

For this question to be well posed, we assume a set of possible extension candidates $\mathcal{E}$ is given. An extension $e \in \mathcal{E}$ can be a new pipe to be built (possibly as a loop) or the insertion of an active element at the beginning or end of an existing pipe in the network. In the case of a new pipe, an active element is always added at one of the end points. This is not only important for our model, but has a practical background. A new pipe connects previously unconnected or only loosely connected parts of the network and might effect the flow and pressure distribution in the entire network. In the extreme case, the construction of a new pipe can render previously feasible nominations infeasible. Closing the active element at the endpoints neutralizes the effect of the pipe and yields the original network.

We assign an integer variable x_e to each extension candidate $e \in \mathcal{E}$. Control valves and compressors augment the functionality of the valve by the respective active state as all elements have closed and open/bypass states. Therefore, when the active state is not used, a much cheaper valve should be built instead of a control valve or a compressor. There are three possible outcomes of the investment decision for extension e which are translated into the variable x_e:

$x_e = 0$: Do not build e.

$x_e = 1$: Build extension e with a valve instead of the proposed active element.

$x_e = 2$: Build extension e with the proposed active element.

If the active element is a valve, then x_e can only take the values 0 and 1. The three options form a hierarchy, where every option is at least as powerful as the ones with a smaller value, but usually at a higher cost.

The general approach consists in extending the network by the candidates and penalizing the use of the extensions by cost on the binary variables of the corresponding active elements. The operation decision of the active element is then translated into the decision whether and how the extension is built. Consider for example a new pipe or loop and its corresponding active element. As closing the active element means that no flow goes through the pipe and the effect of the pipe is neutralized, no cost is associated to the closed state. Using the bypass state means that flow goes along the pipe, but the active element is not used in its active state. Thus, it suffices to build a valve to activate and deactivate the new pipe. The cost for the bypass state is thus the cost for the new pipe plus the cost of a valve. Finally, if the active state is used, then the pipe and the proposed element have to be constructed and therefore the cost associated to the active state is the cost of the pipe plus the cost of the proposed active element. The translation from the operation variables s_a^{bp} and s_a^{ac} to the investment variable x_e is in this case

$$x_e = s_a^{\text{bp}} + 2s_a^{\text{ac}}.$$

In the other case of an active element being added to the end of an existing pipe, the meaning of bypass and closed state are reversed. When in bypass, the proposed element has no effect and no cost occur while when in closed state, a valve has to be constructed:

$$x_e = 1 - s_a^{\text{bp}} + s_a^{\text{ac}}.$$

The cost for the extension e is modeled by an increasing function $c_e(x_e)$ which is only evaluated at integer points. Using the variables s_a^{bp} and s_a^{ac} the objective can be formulated easily.

We denote by $\mathcal{F}$ the set of all vectors $x = (x_e)_{e \in \mathcal{E}}$ such that the extended network allows a feasible operation. In our situation, a closed form description of $\mathcal{F}$ is not at hand and optimization over this set corresponds to the solution of a non-convex MINLP due to the complex model for physics and discrete decisions. In an abstract form, the deterministic extension planning problem can now be stated as

$$\min c(x) \qquad\qquad \text{(SingleScen)}$$
$$\text{s.t. } x \in \mathcal{F}$$

This formulation is complete, but hides the difficulties in describing and optimizing over the set $\mathcal{F}$.

An MINLP model of this problem can be solved by different techniques. The approach used in this work uses an spatial branch-and-bound approach (Section 2.2) with a linear outer approximation of the nonlinear function that is refined in the course of the algorithm. Figure 6.2a shows the feasible set for the equation $|q_a|q_a = z_a$ as solid line and a linear relaxation as shaded area. Figure 6.2b shows the approximations of both subproblems that arise from branching on $q_a = 0$ drawn in the same plot. An alternative is the approximation or relaxation of the nonlinear function by piecewise linear functions which yields a MILP to be solved (see for example [Koc+15]).

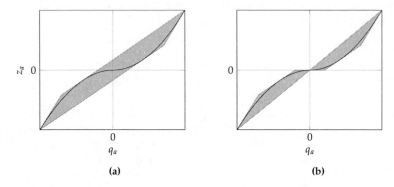

Figure 6.2: Nonlinear equation $|q_a|q_a = z_a$ and a linear outer approximation on the original interval (a) and after one branch at $q_a = 0$ (b).

7 Gas network topology planning for multiple scenarios

The extension of the gas infrastructure involves long-lasting and cost-intensive investments. The operators therefore seek extensions that solve several potential bottleneck situations at once, are flexible in the future operation and play well with possible subsequent network extensions. Clearly, deterministic optimization does not respect any of these objectives. Instead it selects a set of extensions that is tailored towards the particular nomination that might not be relevant for the future. It is therefore of high importance to consider several nominations simultaneously in order to avoid over-tuning and prepare the network for a large range of different flow distribution patterns. Considering uncertainty in the form of a set of scenarios leads to more flexible network extensions that can meet future demands more efficiently.

In this chapter, we present a robust model for gas network extension that protects the TSO against a finite set of scenarios (i.e., transport requests). A branch-and-bound algorithm based on scenario decomposition is proposed that exploits the hierarchical structure of the network extensions. Network extension problems for the individual scenarios are solved as subproblem using the methods presented in the previous chapter. The algorithm provides lower bounds on the obtainable cost such that the quality of solutions can be accessed. While the algorithm is guaranteed to find the optimum solution, we incorporate heuristic methods that prove capable of finding high quality solutions and in turn speed up the optimization. A computational study on realistic network topologies shows the effectiveness of our approach. Our method is able to solve challenging instances with up to 256 scenarios whose scenario-expanded MINLP formulations have hundreds of thousands of variables and constraints to global optimality in a reasonable amount of time.

This chapter is organized as follows. First, we define the robust gas network extension problem in Section 7.1. The decomposition method for a finite uncertainty set is presented in Chapter 7 together with some details about primal and dual bounds and results on the ability to reuse solutions from previous optimization runs over the same scenario. Section 7.6 presents the results of computational experiments. Section 7.7 provides an outlook on planned future work on the topic.

The work presented in this chapter has been performed within the FORNE project[1] while the author was at Zuse Institute Berlin (ZIB). FORNE was a joint research project of Europe's largest gas network operation Open Grid Europe GmbH (OGE) and several academic institutions that ended in 2015. The development of automated methods for gas network extension planning in particular considering uncertainty was one of the project goals. An implementation of our methods has been used by OGE.

A paper coauthored by Frauke Liers has been submitted to the journal "Optimization and Engineering" and is currently under review. A short version of this paper has been published in the OR Proceedings 2014 [Sch16].

7.1 A robust model for gas network extension

Robust Optimization ([BGN09], Section 2.3) is a framework to protect against uncertainty in the input data of an optimization problem. Instead of assuming that the data that describes the objective and the constraints is known, the input data assumed to realize itself within an uncertainty set. The decisions that are to be determined then must be robust, i.e., they need to be feasible no matter how the data manifests itself with the uncertainty set. Furthermore, a robust solution is sought that yields the best guaranteed cost.

For linear mixed-integer problems, tractable robust counterparts can be derived for several classes of uncertainty sets, such as conic or polyhedral sets. As we are facing a complex MINLP, much less is known about tractable robust counterparts. We therefore consider a discrete uncertainty set that consists of a finite number of nominations. This reflects the situation in which different scenarios are collected from historical data, future forecasts, or experts, which was also the case for our industry partner. We will refer to the elements in the uncertainty set as *scenarios*.

The decision variables in our application naturally decompose into two stages: In the first stage, the decision which extensions are built has to be taken. In the second stage, the operational decisions, i.e., the control of the active devices and the resulting physical values such as pressure and flow, have to be determined for all scenarios. While the first stage decisions are taken once for all scenarios, the second stage decisions are taken independently in each scenario and have to take the first stage decisions into account. Accommodating robust multi-stage optimization problems is an active research field and several different approaches

---

[1]http://www.zib.de/projects/forne-research-cooperation-network-optimization

have been proposed, among them *Adjustable Robust Optimization* [Ben+04] and *Recoverable Robust Optimization* [Lie+09].

In the context of our multi-scenario extension planning problem, the scenarios represent nominations and we seek for a set of extensions at minimal cost such that the resulting network allows a feasible operation of *all* scenarios. We stress that in the different scenarios not all extensions that have been built have to be actually used. However, the hierarchical model of network extensions from Section 6.3 ensures that extensions can always be used at a smaller scale.

The problem can then be formulated as a two-stage robust program. The first stage variables y indicate the extent to which extensions are built and are independent of the scenario. In a particular scenario ω, the second stage variables $x^\omega \in \mathcal{F}^\omega$ describe the extent to which the extensions are used. The multi-scenario problem can then be stated as:

$$\min c(y) \qquad \text{(MultiScen)}$$
$$\text{s.t. } x^\omega \in \mathcal{F}^\omega \qquad \text{for all } \omega \in \Omega \qquad (7.1)$$
$$x^\omega \leq y \qquad \text{for all } \omega \in \Omega \qquad (7.2)$$
$$\underline{y} \leq y \leq \overline{y} \qquad (7.3)$$

The constraints (7.2) ensure that an extension used in at least one scenario also has to built. Together with the increasing objective function $c(y)$, (7.2) simply is the linearization of the maximum

$$y = \max_{\omega \in \Omega} x^\omega. \qquad (7.4)$$

Clearly, all variables have to be nonnegative, but we state explicit bounds as they will be handy in the description of the algorithm in Section 7.2. Integrality constraints can be omitted as they are encoded in $\mathcal{F}^\omega$ for x^ω and enforced by (7.4). Note that this model is only valid because the extensions form a hierarchy where more expensive extensions only add functionality. The model also accounts for the fact that extensions might be used to a smaller extent than possible.

In principle, a scenario-expanded problem can be formulated by adding all constraints that describe the relationship $x^\omega \in \mathcal{F}^\omega$ explicitly to the model. Then the operational decisions get another index for the scenario as they act on the second level of the problem. This formulation could be solved by MINLP solver for example by a spatial branching algorithm. However, since the problem is challenging for even one scenario, there is no hope that the resulting MINLP can be solved for a non-trivial number of scenarios.

The simple constraints (7.2) are the only connection between different scenarios. Without these constraints the model would decompose as each scenario problem could be solved individually. A decomposition approach therefore seems most promising for this model. Known decomposition approaches make structural assumptions that are not fulfilled in our model. Classical Generalized Benders Decomposition [Geo72] requires convexity to provide optimal solutions [SG91]. A non-convex variant was described in [LTB11] and was used to solve a stochastic pooling problem for gas network planning under uncertainty [LTB16] but allows binary variables only in the first stage. In general, due to the lack of knowledge of the structure in the set $\mathcal{F}^\omega$, feasibility cuts that carry more information than just forbidding one particular assignment of y are difficult to obtain. We therefore propose a decomposition where the constraints (7.2) are ensured by branching on the y variables.

7.2 Scenario decomposition: A branch-and-bound approach

In the following we outline the algorithmic approach that consists in scenario decomposition in combination with branching on y variables.

First, we solve the scenario subproblems (SingleScen) independently and possibly in parallel for all scenarios $\omega \in \Omega$. Due to the complexity of the problem, which is a non-convex MINLP even for one scenario, we aim to leverage our capabilities of solving these problems and chose a setting where second stage decisions are decided by a black-box solver we don't want to interfere with. This way we also directly benefit from future improvements of solvers for non-convex MINLP. The non-convex problems are encapsulated but we can still use the known structure of the solution space in the design of the algorithm.

If one scenario subproblem is infeasible, the multi-scenario problem is infeasible. If all subproblems are feasible, we denote the best solution found for scenario ω by x^ω. A feasible solution to the multi-scenario problem can be computed by setting

$$y_e^\star = \max_{\omega \in \Omega} x_e^\omega,$$

i.e., by building all extensions that are used in at least one scenario.

Next, we identify extensions that disagree in the extent the extension should be built, i.e., extensions $e \in \mathcal{E}$ for which it is

$$\min_{\omega \in \Omega} x_e^\omega \neq \max_{\omega \in \Omega} x_e^\omega. \tag{7.5}$$

Branching on the y variables is used to synchronize the investment decisions in the different scenarios. To this end, an extension e for which (7.5) holds and a value τ between $\min_{\omega \in \Omega} x_e^\omega + 1$ and $\max_{\omega \in \Omega} x_e^\omega$ is chosen and two subproblems, i.e., nodes in the branch-and-bound tree, are created: one with the condition $y_e \leq \tau - 1$ and one with the condition $y_e \geq \tau$. In the two nodes that emerge the variables y now have non-default bounds, but otherwise the structure of (MultiScen) is unchanged. In consequence, a branch-and-bound tree is built, where each node is identified by the bound vectors $\underline{y}$ and $\overline{y}$.

In the nodes, the subproblems have to be modified in order to reflect the bounds on the y variables. Extensions e whose lower bound $\underline{y}_e$ is greater than zero are built to this extent and the cost are charged as fixed cost in the subproblems. In addition, the extension might be used with a value larger than $\underline{y}_e$, in which case additional cost is charged. The cost is thus computed as $\max\left(c_e(\underline{y}_e), c_e(x_e^\omega)\right)$, an expression which is linearized easily. Upper bounds $\overline{y}_e$ are applied to x_e^ω to control the use of extension e in scenario ω.

The modified single-scenario problem for scenario ω and bounds $\underline{y}$ and $\overline{y}$ then reads as:

$$\min \sum_{e \in \mathcal{E}} \max\left(c_e(\underline{y}_e), c_e(x_e^\omega)\right) \qquad \text{(SingleScen}_\omega\text{)}$$
$$\text{s.t. } x^\omega \in \mathcal{F}^\omega$$
$$x^\omega \leq \overline{y}$$

Hence the subproblem is again a single-scenario extension planning problem with an adapted objective function and upper bounds on some variables which forbid the usage of certain extensions.

In most branch-and-bound algorithms a relaxation is used to guide the algorithm and produce lower bounds. Tighter bounds on the problem remove solutions from the relaxation and thus the objective function of the relaxation can only become worse by tightening the bounds. In our case, the value of the multi-scenario problem (MultiScen) also deteriorates as the bounds get tighter since the search space is restricted. Therefore nodes which are deeper in the tree, i.e., have tighter bounds, have a lower or an equal optimal objective value than higher ones. A lower bound on the solution value of the multi-scenario problem associated to a node of the branch-and-bound tree, i.e., to a pair of bounds $(\underline{y}, \overline{y})$, enables to prune the node if this lower bound is worse, i.e., larger, than the value of the best known solution. In this case, no improving solution can be found in the subtree associated to the node and the node can reliably be pruned from the branch-and-bound tree. A lower bound on the solution value of a minimization problem is also referred to as *dual bound* while feasible solutions are also known as *primal solutions* and the value of the best known feasible solution as *primal bound*. If primal and dual bound coincide, the problem is solved to optimality as the dual bound ensures that no solution with a better objective value can exist. The following two sections study dual bounds and primal solutions for our problem.

7.3 Dual bounds

Lower bounds for the single-scenario problems can be instantly translated into lower bounds for the multi-scenario problem. Intuitively, the cost to ensure simultaneous feasibility of all scenarios has to be greater than for any single scenario. The following short lemma formalizes the argument.

Lemma 7.1: *Let the objective function be non-negative. Then any dual bound for problem* (SingleScen$_\omega$) *for any scenario is also a dual bound for problem* (MultiScen).

Proof. Let $\bar{c}$ be a dual bound to (SingleScen$_\omega$) for scenario ω, i.e., $\bar{c} \leq c(x^\omega)$ for $x^\omega \in \mathcal{F}^\omega$. With constraint (7.2) and the fact that the objective function $c(.)$ is increasing in every direction, we have $\bar{c} \leq c(x^\omega) \leq c(y)$ for any feasible y. Therefore, $\bar{c}$ is also a lower bound for problem (MultiScen). □

It is clear that the constant value $c(\underline{y})$ is a lower bound on the objective value for (SingleScen$_\omega$). As tighter bounds alter the objective function and the solution space of the subproblems it is not obvious that the solution value of the subproblem might only increase with tighter bounds $(\underline{y}, \overline{y})$. The following lemma however states that this is the case.

Lemma 7.2: *Let c^* be the optimal value of* (SingleScen$_\omega$) *for some scenario ω and for the bounds $(\underline{y}, \overline{y})$. Consider a second pair of more restrictive bounds $(\tilde{\underline{y}}, \tilde{\overline{y}})$ with $\tilde{\underline{y}} \geq \underline{y}$ and $\tilde{\overline{y}} \leq \overline{y}$ and its optimal value $\tilde{c}^*$ for* (SingleScen$_\omega$). *Then $c^* \leq \tilde{c}^*$.*

Proof. We use induction over $n = \|\tilde{\underline{y}} - \underline{y}\| + \|\tilde{\overline{y}} - \overline{y}\|$ where $\|.\|$ is the ℓ_1-norm. For $n = 0$, the problems and thus their optimal objective values coincide and our claim holds. For the induction step $n = 1$, we distinguish between a tightened upper and lower bound. In case a lower bound is tightened, i.e., $\tilde{\underline{y}}_e > \underline{y}_e$ for some e, then the search space remains the same, but the objective function increases for $\underline{y}_e = x_e^\omega$. In the case where an upper bound is tightened, i.e., $\tilde{\overline{y}}_e < \overline{y}_e$, the search space is restricted and the objective functions remains unchanged. In both bases the objective function value deteriorates. □

This result will be used in later sections as it ensures consistency of the dual bound of the subproblem.

7.4 Primal solutions

We propose three ways to generate or to improve feasible solutions:

7.4.1 From the solutions of the subproblems

First, by construction the union of all extensions used in the different scenarios constitutes a primal solution for the multi-scenario problem. Therefore, we construct a solution to (MultiScen) in every node by setting $y = \max_{\omega \in \Omega} x_e^\omega$ where x_e^ω is taken as the best solution for scenario ω.

7.4.2 1-opt heuristic

Second, we observed that checking if a small subset of extensions is feasible is typically very fast. This observation is used by a 1-opt procedure that takes a solution to (MultiScen), decreases one variable that has been chosen to take value $y_e > 0$ by 1, and checks all scenarios for feasibility. Of course, a priori it is not clear which $y_e > 0$ to choose. Several options have been explored. The most promising is to sort the y_e according to the possible saving $c_e(y_e) - c_e(y_e - 1)$ realizable by decreasing its value by one and consider extensions with small saving first. The rationale behind this is that often these "small" extensions are used by scenarios which are rather close to feasibility and some more expensive extensions used in the more challenging scenarios often also ensures feasibility of these almost feasible scenarios. Therefore chances are high that these extensions can be removed. Extensions with higher savings are likely to render some scenario infeasible as their effect can't be compensated by the other extensions in the solution.

In order to protect ourselves against outliers, a strict timelimit is used when checking the scenarios for feasibility. Note, that during the 1-opt heuristic all extensions are fixed and the problem is a feasibility problem (nomination validation). Therefore, only a solution is needed to solve the problem and no time is used to prove optimality.

7.4.3 Best-known heuristic

Third, we solve an auxiliary problem to compute the best known multi-scenario solution taking into a account all solutions to the subproblems that the solver found during its solution process.

Optimal single-scenario solutions are very specialized in fixing the bottleneck of the particular scenario. When facing uncertainty in the data, these solutions are not likely to be feasible in the perturbed problems. The optimal solution to the problem with uncertain data is thus often suboptimal for each individual scenario but is able to balance the needs for the different scenarios (e.g., [Wal10]). In this light, it seems reasonable to consider all known solutions to scenarios in order to construct a multi-scenario solution.

During the solution process of a branch-and-bound solver, all feasible solutions the solver encounters are collected and stored in a *solution pool*. These include solutions that are the best known at the time of finding them but also non-improving solutions that might be used in improvement heuristics by the solver. During and in particular after the optimization, the user can query all solutions in the pool. We collect the solutions for some scenario ω in the set $\mathcal{S}^\omega \subseteq \mathcal{F}^\omega$. Using these solutions provides two benefits. First, they might be suitable solutions to one of the single-scenario problems that are solved in the remaining solution process. All solutions from $\mathcal{F}^\omega$ are therefore added to the solver as start solutions whenever a single scenario needs to be solved. Second, we construct the "best known" solution for the multi-scenario solution by solving an auxiliary MILP. To this end, we use indicator variables z^s for each $s \in \mathcal{S}^\omega$ and each scenario ω. To ensure the scenario feasibility constraint (7.1) the MILP has select one solution for each scenario. Furthermore, we need to ensure that all extensions that are used in the selected solutions are built. This leads to the following MILP:

$$\min c(y) \tag{7.6a}$$

$$\text{s.t.} \sum_{s \in \mathcal{S}^\omega} z^s = 1 \qquad \text{for all } \omega \in \Omega \tag{7.6b}$$

$$s \cdot z^s \leq y \qquad \text{for all } s \in \mathcal{S}^\omega, \omega \in \Omega \tag{7.6c}$$

$$z^s \in \{0,1\} \qquad \text{for all } s \in \mathcal{S}^\omega, \omega \in \Omega \tag{7.6d}$$

$$\underline{y} \leq y \leq \overline{y} \tag{7.6e}$$

The program has the building indicator variables y and the same objective function as before. Auxiliary variables to formulate the objective function are omitted. Constraint (7.6b) says that a feasible solution has to be selected for every scenario ω. The term $s \cdot z^s$ in (7.6c) is a vector-scalar multiplication whose result is a (column-)vector as is y. Constraint (7.6c) says, that extensions used in a solution s have to be built if the solution is selected, i.e., $z^s = 1$. The solution to this program gives the best solution to the multi-scenario problem taking all solutions that the solver has found so far for the single scenarios into account.

Note that the program (7.6) is also a complete description of the multi-scenario problem if $\mathcal{S}^\omega = \mathcal{F}^\omega$, i.e., if $\mathcal{S}^\omega$ is the set of all feasible solutions. In [SPW09] this formulation is described and the authors propose to perform Dantzig-Wolfe decomposition to solve the continuous relaxation of (7.6). The pricing problem is then again a single-scenario problem with the dual variables in the objective function. In their application these solutions are mostly integer feasible. In principle heuristics or a Branch-and-price scheme is needed to generate integer feasible solutions with this approach.

7.5 Reusing solutions

During the course of the proposed branch-and-bound procedure, the bounds $\underline{y}$ and $\overline{y}$ are tightened, and adjusted single-scenario problems are repeatedly solved. Two instances of the single-scenario problem for a scenario only differ in the objective function and the upper bound on the extension variables as discussed in Section 7.2. In some important cases, not all scenario subproblems need to be solved again since we already know the optimal solution. As an example, take the extreme case where a scenario is found to be feasible without extensions. Clearly, the procedure should never touch this scenario again.

In order to decide whether a solution $S \in \mathcal{F}^\omega$ from a previous node can be reused, we need to take the bounds $(\underline{y}^S, \overline{y}^S)$ under which the solution was computed and the current bounds $(\underline{y}, \overline{y})$ into account.

We start with the simple observation that if all the extensions in a solution are already built, then the solution is optimal for the restricted problem:

Lemma 7.3: *If $S \in \mathcal{F}^\omega$ and $S \leq \underline{y}$, then S is an optimal solution for (SingleScen$_\omega$) for bounds $(\underline{y}, \overline{y})$.*

Proof. Clearly S is feasible for (SingleScen$_\omega$) for bounds $(\underline{y}, \overline{y})$. As the objective function is increasing, its cost equals $c(\underline{y})$ which is a lower bound for the subproblem. □

Observe that in this case it is irrelevant for which bounds S was computed and that it is not required that it was an optimal solution when it was computed.

The previous lemma examined the situation where all extensions that are used are already built. The next lemma treats the opposite situation where we are using more than has been built and no built extension is unused. In this situation the solution has to be optimal for some bounds and the new bounds need to be stronger than the previous ones.

Lemma 7.4: *Let $S \in \mathcal{F}^\omega$ be an optimal solution to (SingleScen$_\omega$) given the bounds $(\underline{y}^S, \overline{y}^S)$. Let $(\underline{y}, \overline{y})$ be tighter than $(\underline{y}^S, \overline{y}^S)$, i.e., $\underline{y} \geq \underline{y}^S$ and $\overline{y} \leq \overline{y}^S$.*
If $\underline{y} \leq S$ and $S \leq \overline{y}$, then S is an optimal solution to (SingleScen$_\omega$) for bounds $(\underline{y}, \overline{y})$.

Proof. The crucial point is the optimality of S given the bounds $(\underline{y}^S, \overline{y}^S)$. Since $S \leq \overline{y}$, the solution is still feasible for $(\underline{y}, \overline{y})$. Since $\underline{y} \leq S$ the solution value of

S remains the same for $(\underline{y}, \overline{y})$ as it was for $(\underline{y}^S, \overline{y}^S)$. The value $c(S)$ was a dual bound to (SingleScen$_\omega$) with respect to the bounds $(\underline{y}^S, \overline{y}^S)$. Due to Lemma 7.2, the dual bound can only have increased by using tighter bounds. In total, S is a feasible solution whose objective value matches the dual bound and is therefore optimal. $\qquad\square$

The previous lemmas dealt with extreme cases where solution dominates the lower bound or vice versa. The situation becomes tricky if a solution S neither dominates nor is dominated by the current lower bound vector $\underline{y}$. In this case the solution does not fully use extensions that are already built but uses extensions that are still undecided. We have to make sure that these unused, but built extensions cannot help to find cheaper overall solution. Define for some solution $S \in \mathcal{F}^\omega$ the set I of extensions where the lower bound has been increased compared to when the solution was found

$$I = \{e \in \mathcal{E} \mid \underline{y}_e > \underline{y}_e^S\}$$

and the set J of those extensions where the solution S uses more than what is already built

$$J = \{e \in \mathcal{E} \mid \underline{y}_e \leq S_e\}.$$

The following lemma generalizes Lemma 7.4.

Lemma 7.5: *Let $S \in \mathcal{F}^\omega$ be an optimal solution given the bounds $(\underline{y}^S, \overline{y}^S)$ and let $(\underline{y}, \overline{y})$ be tighter than $(\underline{y}^S, \overline{y}^S)$, i.e., $\underline{y} \geq \underline{y}^S$ and $\overline{y} \leq \overline{y}^S$.*
If $I \subset J$ and $S \leq \overline{y}$, then S is an optimal solution to (SingleScen$_\omega$) for bounds $(\underline{y}, \overline{y})$.

Proof. In contrast to Lemma 7.4, there might be extensions that are decided to be built to a larger extent than they are used in the solution ($\underline{y}_e > S_e$). However, the amount by which extensions have been built additionally, i.e., where the lower bound increased, does not suffice to exceed the usage of the extensions. Hence, the objective function value for S is identical for both sets of bounds. Since the dual bounds can only increase by tightening the bounds and since the solution is optimal for the bounds $(\underline{y}^S, \overline{y}^S)$, the solution is still optimal given the bounds $(\underline{y}, \overline{y})$. $\qquad\square$

To illustrate the usefulness of the previous lemmas, consider the situation after branching for the first time. At the root, we assume all scenarios are solved to optimality and a branching point τ is chosen for some extension e. In the first branch, the constraint $y_e \leq \tau - 1$ is added. Clearly, all scenarios ω with $x_e^\omega \geq \tau$ in the optimal solution x need to be solved in this branch because their optimal solution has been cut off and the dual bound for these scenarios might increase due to the additional constraint. For those scenarios ω with $x_e^\omega \leq \tau$, Lemma 7.4 holds and we know that optimal solutions computed in the root remain optimal in this branch. In the second branch, the constraint $y_e \geq \tau$ is added and the lower bound is tightened. All scenarios with $x_e^\omega < \tau$ have to be solved again, unless they fulfill the conditions of Lemma 7.3 which in this case means that $S = \underline{y}^S$. Optimal solutions of those scenarios with $x_e^\omega \geq \tau$ again fulfill the conditions of Lemma 7.4 and those scenarios don't have to be solved again.

7.6 Computational Experiments

To show that our approach can solve practical problems we conducted extensive computational experiments an realistic gas network topologies. Our subproblem is a non-convex MINLP and topology optimization for even one scenario is a big challenge; even deciding feasibility of a nomination is a difficult task, see for example [Koc+15] where the nomination validation problem has been extensively studied and it has been shown that problems which are on the boarder between being feasible and infeasible are particularly challenging. In this situation, the linear relaxation is often feasible and it needs a lot of spatial branching to proof infeasibility. The situation where a set of extensions almost suffices to ensure feasibility of a scenario and a large effort of spatial branching has to be made to proof this is not the case is expected to occur frequently during our algorithm.

Nevertheless, we are able to provide optimal solutions for instances with up to 256 scenarios whose deterministic equivalent problems have almost 200 000 variables and over 225 000 constraints from which 80 000 are nonlinear. Furthermore, we provide feasible solutions with a proven optimality gap of 16 % for an instances whose deterministic equivalent has more than 360 000 variables and 220 000 constraints from which more than 68 000 are nonlinear.

7.6.1 Computational Setup

The approach is implemented using the MINLP solver SCIP [Ach09; VG16] in the framework Lamatto++ [GMM14], which is used for data handling. SCIP provides the core of the branch-and-bound algorithm. A proper constraint handler ensures the abstract constraint (7.1) and is able to solve single-scenario problems. Methods to solve the single-scenario problems were developed in the FORNE project. A relaxator plugin triggers the solution of the single-scenario problems, returns dual bounds to SCIP and identifies branching candidates. The heuristics from Section 7.4 are implemented as heuristic plugins. SCIP is also used to solve the

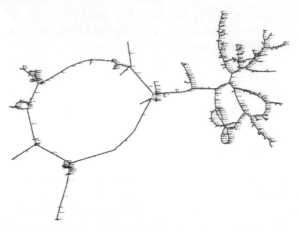

Figure 7.1: Visualization of the gaslib-582 network. Source: [Gasb] License: [Cre]

MILP (7.6). SCIP is used in a development version (shortly before the 3.1 release) and calls CPLEX version 12.5.1 to solve LP-relaxations.

7.6.2 Testsets and Instances

To test our approach, we used the networks of the publicly available gaslib [Gasa; Hum+15] under the Creative Commons Attribution 3.0 Unported License [Cre]. The gaslib contains three networks of different sizes. The biggest one, gaslib-582, is a distorted version of real data from the German gas network operator Open Grid Europe GmbH and comes with a large number of nominations. The smaller ones gaslib-40 and gaslib-135 are abstractions of the first. They are given mainly for testing purposes and come with one reference scenario. For gaslib-582 we used the associated nominations and scaled them to simulate growing demand which renders increasing infeasibilities with the original network. For gaslib-40 we were able to generate scenarios and extensions that yield promising instances. For gaslib-135, however, the same algorithm did not produce instances where SCIP could find feasible solutions for the single scenarios within the timelimit for subproblems. We therefore skip this network and use only the largest gaslib-582 and the smallest gaslib-40 networks for our computational study.

The generation of meaningful scenarios and extension candidates is a difficult task. The scenarios should be infeasible and have different bottlenecks that can be fixed by a large number of different extensions from the candidate set

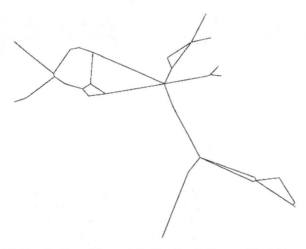

Figure 7.2: Visualization of the `gaslib-40` network. Source: [Gasc] License: [Cre]

in order to yield interesting instances for a robust planning approach. As our approach solves the single-scenario problems repeatedly, it is important to select nominations that are known to be feasible. Infeasible nomination directly render the multi-scenario problem also infeasible while nominations where no feasible solution for the topology planning problem can be found will block our approach. Both situations are not in the interest to study the behavior of our approach. In the following we describe the instances on the two networks in more detail.

`gaslib-582:` To test we approach, we seek a large number of infeasible scenarios in order to be able to construct a reasonable number of instances with multiple scenarios. However, feasible solution to the single-scenario problems should be found within the timelimit.

As a first step, a list of possible network extensions has to be created. The generation of meaningful extension candidates is an art by itself. Considering that every possible network extension adds binary decisions and in the case of pipes also nonlinearities to the problem, adding too many candidates renders the selection of an optimal subset for a single scenario impossible. Too few extensions on the other hand might not allow a feasible solution for all scenarios and are to inflexible for challenging multi-scenario instances.

For the generation of network extensions we relied on techniques developed in the FoRNE project. For the `gaslib-582` network, we used two methods to create extensions. First, for one of the scenarios all pipes where the flow is fixed by a simple flow propagation are looped. The rationale is that if the flow is fixed, then difference of the squared pressures is also known and the nonlinearities describing the physics of this pipe and its loop disappear. Computationally, these loops are expected to be very cheap. The second method proposes new pipes by evaluating the reduction in the transport momentum caused by adding a pipe. The transport momentum for a scenario is computed by disregarding the gas physics and considering only a linear flow problem. The objective is to minimize the sum over all pipes of the product of the length of the pipe with the flow through it. It is a measure how efficiently flow can be routed through the network. New pipes whose endpoints are far enough apart and whose addition to the network results in the largest reductions of the transport momentum are selected as extension candidates. Meaningful geographic coordinates (or a meaningful distance matrix) are essential for the computation of the transport momentum and we therefore consider this approach especially useful for the realistic gaslib-582 network.

To ensure we have a sufficiently large number of infeasible scenarios, all input and output flows have been scaled by a factor of 1.2. Then, we performed a single-scenario topology optimization to filter out those instances that are still feasible without network extension and those that don't find a solution with the timelimit for subproblems. The result are 107 scenarios that exhibit a positive objective value after 600 seconds.

It is worthwhile noting that this procedure is non-deterministic because of the timelimit used. The path taken to solve the problem is deterministic, but if the solution is found very close to the timelimit, a random disturbance might cause a slowdown and the solution might not be found within the timelimit in the next run. It therefore can (and actually does) happen that in a multi-scenario run some scenario does not find a solution in the root node. A limit based on the number of simplex iterations or number of branch-and-bound nodes would eliminate this problem, but is not practical in our application.

In the next step, the 107 scenarios have been grouped together to construct multi-scenario instances. The aim was to construct a set of instances and to make sure that all scenarios participate in the mix. The procedure was to first shuffle the list of scenarios. Then, assuming that k is the desired number of scenarios in an instance, the first k scenarios in the list are selected and removed from the list. This is repeated until the length of the list of unused scenarios is smaller than k. Then the list of all scenarios is shuffled again and the procedure is repeated. We composed 50 instances of 4, 8, and 16 scenarios each, 25 instances with 32 scenarios and 10 instances with 64 scenarios. Finally, we add one instance with

Scenarios	Variables		Constraints			
	Total	Binary	Total	Linear	SignPower	Indicator
gaslib-582						
4	13 740	624	12 336	8 296	2 568	1 472
8	27 480	1 248	24 672	16 592	5 136	2 944
16	54 960	2 496	49 344	33 184	10 272	5 888
32	109 920	4 992	98 688	66 368	20 544	11 776
64	219 840	9 984	197 376	132 736	41 088	23 552
107	367 545	16 692	329 988	221 918	68 694	39 376
gaslib-40						
4	3 044	168	3 524	1 948	1 256	320
8	6 088	336	7 048	3 896	2 512	640
16	12 176	672	14 096	7 792	5 024	1 280
32	24 352	1 344	28 192	15 584	10 048	2 560
64	48 704	2 688	56 384	31 168	20 096	5 120
128	97 408	5 376	112 768	62 336	40 192	10 240
256	194 816	10 752	225 536	124 672	80 384	20 480

Table 7.1: Size of the deterministic equivalent formulations

all 107 scenarios.

For the `gaslib-582` testset, the subproblems are solved in parallel with up to 16 threads whenever possible. We used a time limit of 600 seconds for the subproblems which is reduced to 60 in the 1-opt heuristic. The total timelimit was set to 12 hours.

Table 7.1 summarized the size of the scenario-expanded formulations of the resulting multi-scenario instances. The table is split into two parts, each for one network topology. Each line contains the number of variables and constraints for the deterministic equivalent for the number of scenarios that is given in the first column. We report the total number of variables and the number of binary variables. For the constraints, we report the total number of constraints and break them down into the three relevant classes for our problem which are linear constraints, SignPower constraints, i.e., constraints of type (6.2), and indicator constraints which are used to model implications such as (6.4)-(6.8). The largest problems for this network have 360 000 variables and 220 000 constraints from which almost 70 000 are nonlinear. These numbers are given for reference purpose only as our decomposition approach does not work on the scenario-expanded formulation. Nevertheless, our approach assigns values to each of the variables and provides guarantees on the quality of the full solution to the multi-stage robust optimization problem.

gaslib-40: In contrast to the gaslib-582 network which contains 4227 realistic nominations, the gaslib-40 network only has one nomination. This scenario is quite artificial as it evenly distributes the flow amount over all 3 entries and 29 exits. The motivation to work with this network however is not primarily the practical relevance of the instances but to challenge our approach using a network where the subproblems are easier to solve than on the more realistic gaslib-582.

We created 2000 nominations using the following algorithm. First, we sampled the number of entries/exists that should have inflow/outflow uniformly between 1 and the number of entries/exists. Then, we randomly picked that assigned number of entries/exits. The total flow from the reference scenario is scaled with a uniformly sampled factor between 0.75 and 2. The resulting scaled total flow is then uniformly distributed among the selected entries and exists, respectively.

For this network, only loops are considered. As before, all pipes where the flow after preprocessing of the reference scenario is fixed can be looped. In addition all pipes that are longer than 20 km are in the candidate set.

From the 2000 nominations, of course a large number is feasible in the original network or still infeasible even with the proposed loops. We therefore proceed as before and solve the single-scenario problem with a timelimit of 600 seconds. In order to keep the time to solve the subproblems as small as possible, we select those nominations that are solved to optimality within the timelimit of 600 seconds that have a positive objective function value. We note that many different loops are used in these solutions such that challenging multi-scenario instances can be expected. The resulting 425 nominations are then grouped into 50 instances of 4, 8, and 16 scenarios each, 25 instances with 32 scenarios and 10 instances with 64, 128, and 256 scenarios each.

For gaslib-40 the parallel solution of single-scenario problems caused buffer overflow errors in the CppAD package for algorithmic differentiation used with SCIP. As these errors are out of our control, these instances are solved purely sequentially which avoids this error.

We used the same time limits of 600 seconds for the subproblems, 60 seconds for subproblems within the 1-opt heuristic. For the overall algorithm we first used 12 hours for all numbers of scenarios on the gaslib-582. As this timelimit is found to be short for 64 and more scenarios, we ran the instances with 64, 128, and 256 scenarios also with a timelimit of 48 hours.

Scenarios	Instances	Opt.	MsaS	Union	LTL	Nodes		Gap [%]
						Finished	All	
gaslib-582								
4	50	28	20	14	20	3.6	16.4	42.8
8	50	21	9	7	26	6.3	24.4	41.9
16	50	8	0	0	34	7.2	36.9	27.3
32	25	1	0	0	16	7.0	47.2	17.4
64	10	0	0	0	6	–	56.5	23.2
107	1	0	0	0	1	–	48.0	16.4
gaslib-40 **(Timelimit 12 hours)**								
4	50	49	12	22	0	13.7	14.2	44.3
8	50	43	7	9	0	24.8	29.7	12.3
16	50	33	1	2	0	35.7	38.5	26.4
32	25	11	0	0	1	27.4	33.2	24.6
64	10	3	0	0	0	23.7	25.4	28.5
128	10	1	0	0	0	67.0	16.1	64.1
256	10	0	0	0	0	–	4.5	98.6
gaslib-40 **(Timelimit 48 hours)**								
64	10	8	0	0	0	56.8	64.1	25.7
128	10	2	0	0	2	80.0	60.7	26.5
256	10	1	0	0	0	27.0	21.2	49.3

Table 7.2: Summary of computational results

7.6.3 Results

Table 7.2 summarizes the performance of our approach. The table is divided into three parts; the first part for results on the gaslib-582 testset and then two parts for results on the gaslib-40 testset with timelimit 12 and 48 hours. The rows are grouped by the number of scenarios considered in each instance. The first two columns then report the number of scenarios and instances in the respective group. The third column gives the number of instances solved to optimality within the timelimit. The next two columns analyze the structure of the best solution found by our approach and compare it with the best solutions known for each scenario. For gaslib-40 the optimal solutions to the single scenario runs are known. For gaslib-582 the best solution after solving the scenario for 12 hours is used as the best known solution for the scenarios. The column *MsaS* states the number of instances where the best solution to the multi-scenario problem builds the same extensions as the best known solution to one of the scenarios. In this case one scenario dominates the others as the extensions needed by the scenario suffice to ensure feasibility of all other scenario problems as well. The column *Union* in contrast brings light into the opposite case where the solution of the multi-scenario problem contains the union of the extensions built in the single scenarios. An instance appears in both columns MsaS and Union if all scenarios select the same extensions. Eventually, the branch-and-bound algorithm comes to the point when all y variables are fixed and the subproblems only have to decide feasibility given the fixed extensions. In this case, the remainder of the global timelimit might be used for the feasibility problem as otherwise the algorithm can't proceed. The column *LTL* states how many instances did not finish to optimality because the remaining time was used in the subproblems and the algorithm was stuck. The last two groups of columns show the average number of nodes, split by the instances that were solved to optimality and all instances, and the average gap of those instances that were not solved to optimality.

While our approach is able to solve 28 out of 50 instances, or 56 % of the instances, with 4 scenarios on the realistic `gaslib-582`, this percentage decreases with increasing number of scenarios. At the same time the number of instances where the algorithm gets stuck because one of the subproblems can't properly decide feasibility is constantly high. At the maximum more than 11 of the 12 hours are spent trying to decide the feasibility of one scenario. On this testset there is also a remarkable difference between the number of nodes of those instances that could be solved to optimality, where the average number is at most 7.2 nodes for 16 scenarios, to those instances that hit the timelimit, where the number goes up to 56.5 for 64 instances. This shows that the instances that could be solved don't need much branching in order synchronize the scenarios and that our heuristics do a good job in finding the optimal solution. The high numbers over all instances are because at many nodes some scenarios don't find a feasible solution, but also don't prove infeasibility or even provide good bounds. In this case, our branching mechanism branches on some unfixed variable. In general, the number of nodes is quite low compared to what we are used to from branch-and-bound MILP or MINLP solvers. This shows that the solutions of the single scenarios provide good indications for the structure of multi-scenario solution, even though their solution is rather time consuming. Also good solutions are found very early in the tree. The primal bound makes pruning and propagation very effective, especially as solutions can typically use only very few extensions because otherwise the cost is higher than the dual bound.

The average gap values reported on the `gaslib-582` are quite satisfactory. Note that the gap is computed as the average of only those instances t hat are not solved to optimality and again we have to see them in the light of the difficulty of the problem. Particularly, an average gap of 23.2 % on the instances with 64 scenarios and 16.4 % gap on the instance with all 107 scenarios shows that the solutions are of high quality. Overall, the ability to provide bounds on the solution quality and, if possible, a certificate for optimality is an advantage of our approach.

High numbers in the MsaS and Union columns indicate that the structure of the optimal solution is such that a manual approach might find a good or even the optimal solution. In this case, either one scenario dominates the solution of the solution consists of the union of all the built extensions in the single scenarios; a situation that is easily recognized in a manual fashion and which renders a more sophisticated approach unnecessary. On both testsets, many optimal solutions to the instances with 4 and 8 scenarios have such a structure. In the instances with 4 scenarios 30 instances on gaslib-582 and 27 gaslib-40 are either dominated by one scenario or the multi-scenario solution is the union of the extensions in the best single-scenario solutions. We note that both can also happen simultaneously. The number goes down to 15 and 11 for gaslib-582 and gaslib-40, respectively, when 8 scenarios are considered and completely disappear for higher number of scenarios except for 3 instances with 16 scenarios on the gaslib-40 network. This shows that for a few scenarios only a manual planning approach based on solving the single-scenarios could provide good or even optimal solutions. For larger numbers of scenarios the manual approaches are unlikely to find good solutions as there the synchronization between the scenarios becomes more important. Of course manual planning approaches also lack quality guaranties in terms of gap to the best possible solution which our approach provides.

On the smaller gaslib-40, all but one instances with 4 scenarios can be solved to optimality. Then the percentage of instances solved to optimality decreases, but still 3 out of 10 instances with 64 scenarios are solved within the timelimit. Still one instances with 128 scenarios is solved, but we observe a strong decrease in the number of nodes processed which indicates that the timelimit is very short for these large numbers of scenarios. Note that on gaslib-40 on average 24 nodes are used in the instances that are solved to optimality, but with 128 and 256 only 16.1 and 4.5 nodes are processed on average, respectively. The large average gaps in these groups of instances then also do not surprise. When increasing the timelimit from 12 to 48 hours many more nodes are processed and 8 out of 10 instances with 64 scenarios, 2 with 128 and 1 instance with 256 scenarios are solved to optimality within the increased timelimit. Also the average gap is reduced considerably giving with 25.7 %, 26.5 %, and 49.3 % for 64, 128, and 256 scenarios, respectively, very reasonable results. On this testset, as intended the

165

Scen.	Inst.	Sols	Subprob				1opt				Best Known			
			Succ	Best	Sols	Time	Succ	Best	Sols	Time	Succ	Best	Sols	Time
gaslib-582														
4	50	3.2	50	7	1.1	2.0	42	39	2.4	144.8	6	2	1.0	0.2
8	50	4.7	50	13	1.4	4.8	45	29	3.3	399.2	11	4	1.1	1.0
16	50	6.0	50	13	2.1	11.7	47	28	3.6	940.1	24	8	1.1	4.1
32	25	8.0	25	5	2.4	27.1	22	8	5.1	2194.6	17	12	1.6	17.4
64	10	6.5	10	7	2.6	57.3	9	0	3.1	2236.2	6	3	1.8	34.7
107	1	6.0	1	1	2.0	103.9	1	0	1.0	2094.8	1	0	3.0	152.3
gaslib-40 (Timelimit 12 hours)														
4	50	3.0	49	24	2.3	0.4	17	13	1.3	24.6	18	12	1.0	0.2
8	50	6.5	50	10	4.7	0.7	33	29	1.9	64.5	29	11	1.0	0.8
16	50	10.5	50	4	6.5	1.5	43	33	3.3	345.8	41	11	1.3	2.5
32	25	14.3	25	2	8.0	3.0	24	21	5.4	951.4	22	2	1.2	4.6
64	10	13.0	10	0	6.1	5.5	10	8	5.3	1336.2	10	2	1.6	11.1
128	10	15.3	10	0	7.4	12.2	9	5	6.3	4377.3	10	5	2.2	26.8
256	10	10.6	10	0	3.5	22.2	10	8	5.5	10174.9	10	2	1.6	36.5
gaslib-40 (Timelimit 48 hours)														
64	10	20.0	10	1	11.2	6.9	10	7	7.1	1935.3	10	1	1.6	17.9
128	10	21.7	10	0	13.2	12.5	9	6	6.7	5782.0	10	4	2.5	44.0
256	10	14.4	10	0	6.5	22.3	10	5	6.0	10732.0	10	5	1.9	52.3

Table 7.3: Statistics about solutions found by the different parts of the algorithm

subproblems can be solved much more reliably and the algorithm is stuck only on one instances where feasibility of the subproblem can't be decided in more than 10 hours.

Table 7.3 analyses the components of the algorithm that produce primal solutions. The structure of Table 7.3 is similar to that of Table 7.2. The column *Sols* states the average number of solutions that have been found in the instances of the respective group. Then three blocks analyze the heuristic components of the algorithm. In each block, we report the number of instances where the component found at least one solution and where it found the best solution (columns *Succ* and *Best*, respectively), the average number of solutions found (column *Sols*), and the average time spent in the heuristic (column *Time*). The first block with header *Subprob* belongs to the solution that is derived by building all extensions that are used in the best solutions of the scenarios, i.e., by setting $y = \max_{\omega \in \Omega} x_e^\omega$. This approach finds solutions for all instances (in one instance which is not marked as success, all scenarios use exactly the same extensions and thus the heuristic is not called as the relaxation already found the optimal solution). The second block *1opt* belongs to the highly effective 1opt heuristic.

Even though it can be time-consuming, it finds plenty of solutions which often constitute big improvements. It is also able to find the best known solutions for a large number of instances. The last block *Best Known* corresponds to the approach where the best known solution is computed by the auxiliary MILP (7.6). This approach is also successful on a broad range of instances and in particular on the most difficult instances with larger numbers of scenarios where it often finds the best solution. The short running times show that the MILP is solved without problems. Overall, we conclude that all proposed heuristics constitute to the success of the algorithm.

The appendix of the online version of this thesis [Sch17] provides detailed performance figures for all instances.

7.7 Conclusion

We presented a method for gas network planning with multiple demand scenarios. The computational experiments show that our approach can provide good solutions with reasonable quality guarantees on realistic network topologies. A large range of instances is solved to proven optimality.

Even though developed in the context of gas network planning, the limited assumptions on the underlying problem structure suggest the generalization to other capacity planning problems in the future. Recall that we only assume that the extensions form a hierarchy where higher levels, i.e., more expensive extensions, have all the functionality of all lower levels and the availability of a black box solver for the adjusted single-scenario problems (SingleScen$_\omega$). In [SPW09], for example, the authors use the same framework and Dantzig-Wolfe decomposition on model (7.6) to approach a rather generic capacity expansion problem.

While it is an advantage of our approach that it assumes no particular structure in the subproblems, the algorithm can be enhanced by using more information about the solution space of the subproblems. For gas networks without active devices in the original network and with only loops as extensions candidates, [Hum14] describes inequalities that enforce that a certain amount of loops has to be built in order to make a scenario feasible. Using inequalities of this type that are found during the solution of the subproblems to propagate bounds on the y variables or to steer the search could be promising way to improve the algorithm in this special application and is subject to future research.

8 Conclusion

In this thesis, we focused on applications, where nonconvexity in the model formulation and uncertainty in the data pose principal difficulties. In each case, we identified structural particularities which we then exploited in the development of new or the improvement of existing algorithms. We conducted extensive computational experiments in order to fathom the computational impact. We summarize our achievements as follows:

- We contributed to version 12.7.0 of the commercial solver CPLEX. We found a practical way to project out linearization variables from inequalities obtained from the Reformulation-Linearization-Technique. An implementation of this is implemented into CPLEX and enabled by default for problems with non-convex quadratic objective. On the CPLEX testset at IBM comprising several hundred problems projected RLT is instrumental for solving eleven additional instances in comparison to version 12.6.3. When looking at models that are affected by the separation of theses inequalities, a runtime reduction of 29 % is achieved. This rises to 84 % when focusing on "hard instances" where either of the two versions takes at least 1000 seconds to solve the problem. (Section 3.3.1)

- We established new inequalities, named Motzkin-Straus Clique and generalized MSC bipartite inequalities, for Standard Quadratic Programs (SQP) and Quadratic Knapsack problems. We presented separation methods and demonstrated their effectiveness in computational experiments. On a testset comprising 240 SQP instances, MSC bipartite inequalities close over 68 % of the remaining gap of the CPLEX root relaxation including the application of RLT. This goes up to 86 % for generalized MSC bipartite inequalities, but at a higher computational cost. With reference to CPLEX version 12.6.3, combining the strengths of projected RLT and both classes of inequalities, we increased the number of instances solved from 119 to 230. The runtime as well as the number of branch-and-bound nodes fall by one or two orders of magnitude depending on the problem size. On a Quadratic Knapsack testset our inequalities close 85 % of the gap to the CPLEX root relaxation and 40 % of the gap to the root relaxation including RLT. (Chapter 4)

- We introduced a tight relaxation for the pooling problem, an important non-convex quadratic program with application, for example, in the petrochemical industry. Based on the convex hull of this non-convex relaxation, we obtained strong valid convex inequalities. Applying structural insights allows to reduce the gap of the state-of-the-art pq-relaxation from 5.5 % to 3.0 % on a testset of 360 instances. Moreover, twelve additional instances are solved to optimality within the timelimit. (Chapter 5)

- We addressed the challenge of extending a large gas transport network under uncertainty – posed by Europe's largest gas network operator Open Grid Europe. We modeled uncertainty by a discrete uncertainty set in a robust optimization setting and presented a tailored decomposition algorithm for the resulting MINLP using the hierarchical structure of network extensions. We carried out computational studies on instances from the public gaslib [Gasa] with up to 256 scenarios, which were not tackled before. Our approach proves capable of providing primal and dual bounds, solving many of them to global optimality. The approach can be generalized to problems with a similar structure. (Chapter 7)

In conclusion, different techniques were used to achieve these results. Standard quadratic programs and the pooling problem are considered in the context of a spatial branch-and-bound algorithm as solution method. For both applications we use the structure of the feasible set to improve the relaxation, but in very different ways. For standard quadratic programming, we use a connection to a combinatorial problem to derive cutting planes that were then generalized using a connection to RLT. In contrast, for the pooling problem, the key contribution was to find and study an appropriate relaxation to derive valid inequalities. For robust gas network planning, a custom decomposition algorithm was developed that uses the hierarchical nature of the network extensions and the decomposability due the the consideration of a finite set of scenarios.

The results of this work can be divided into two groups:

- *Detect and exploit specific substructures in complex models to be solved by general-purpose solvers.* Following this approach, software packages for problem classes such as LP, MILP but also MINLP apply a "bag of tricks" [Bix+] and solve many practical problems very efficiently. Motzkin-Straus Clique inequalities and generalized MSC bipartite inequalities, that are based on easily detectable structures, as well as projected RLT inequalities add tricks to the bag for a large variety of non-convex quadratic problems.

- *Exploit the special structure of models from a specific application to develop tailored, special-purpose algorithms.* This approach powered great success in the practical solution of problems from numerous applications that are out of reach for today's general-purpose solvers. Examples include the successful work in the FORNE project in general and our work on the robust gas network expansion problem in particular. The findings on strong relaxations for the pooling problem also stem from an in-depth study of the specific application, but might become available to general-purpose MINLP solvers as methods to detect pooling substructures in complex models are becoming available [CKM16].

Our ambition is to exploit the mathematical structure of problems to computationally efficiently solve them. We found that isolating these structures in stylized problems like Standard Quadratic Programs or the Pooling Problem fosters mathematical insights, where real-world instances are overcharged with details which obfuscate the relevant underlying structures. Exploiting structural knowledge from specific applications in general-purpose solvers is another challenge and we believe this is an essential part of evolving computational MINLP into the well-adopted, stable, and versatile tool that MILP already is.

Bibliography

[AAF98]   C. S. Adjiman, I. Androulakis, and C. Floudas. "A Global Optimization Method, αBB, for General Twice-Differentiable Constrained NLPs - II. Implementation and Computational Results." In: *Computers and Chemical Engineering* 22 (1998), pages 1159–1179 (cited on pages 16, 23).

[Ach09]   T. Achterberg. "SCIP: Solving Constraint Integer Programs." In: *Mathematical Programming Computation* 1.1 (2009), pages 1–41 (cited on page 157).

[Adj+96] C. S. Adjiman, I. Androulakis, C. Maranas, and C. Floudas. "A Global Optimization Method αBB for Process Design." In: *Computers and Chemical Engineering* 20 (1996), pages 419–424 (cited on page 23).

[Adj+98] C. Adjiman, S. Dallwig, C. A. Floudas, and A. Neumaier. "A Global Optimization Method, αBB, for General Twice-Differentiable Constrained NLPs - I. Theoretical Advances." In: *Computers and Chemical Engineering* 22 (1998), pages 1137–1158 (cited on pages 16, 23).

[AE08]   H. Almutairi and S. Elhedhli. "A new Lagrangean approach to the pooling problem." In: *Journal of Global Optimization* 45.2 (2008), pages 237–257 (cited on page 87).

[AF83]   F. A. Al-Khayyal and J. E. Falk. "Jointly Constrained Biconvex Programming." In: *Mathematics of Operations Research* 8 (1983), pages 273–286 (cited on pages 25, 116).

[AH13a]   M. Alfaki and D. Haugland. "Strong formulation for the pooling problem." In: *Journal of Global Optimization* 56 (2013), pages 897–916 (cited on pages 86, 87).

[AH13b]   M. Alfaki and D. Haugland. "A multi-commodity flow formulation for the generalized pooling problem." In: *Journal of Global Optimization* 56.3 (2013), pages 917–937 (cited on page 87).

[AH14]   M. Alfaki and D. Haugland. "A cost minimization heuristic for the pooling problem." In: *Annals of Operations Research* 222.1 (2014), pages 73–87 (cited on pages 17, 87).

[AMF95]   I. P. Androulakis, C. D. Maranas, and C. A. Floudas. "αBB : A Global Optimization Method for General Constrained Nonconvex Problems." In: *Journal of Global Optimization* 7 (1995), pages 337–363 (cited on pages 16, 23).

[Ans12]   K. M. Anstreicher. "On convex relaxations for quadratically constrained quadratic programming." In: *Mathematical Programming* 136.2 (2012), pages 233–251 (cited on page 23).

[ANT]   ANTIGONE. URL: http://helios.princeton.edu/ANTIGONE (cited on pages 1, 17).

[ATS99]   N. Adhya, M. Tawarmalani, and N. V. Sahinidis. "A Lagrangian Approach to the Pooling Problem." In: *Industrial & Engineering Chemistry Research* 38.5 (1999), pages 1956–1972 (cited on pages 87, 131).

[Aud+04] C. Audet, J. Brimberg, P. Hansen, S. L. Digabel, and N. Mladenovic. "Pooling problem: Alternate formulations and solution methods." In: *Management Science* 50 (2004), pages 761–776 (cited on pages 17, 87, 131).

[BAR]   BARON. URL: http://minlp.com/baron (cited on pages 1, 17).

[BBC11]   D. Bertsimas, D. B. Brown, and C. Caramanis. "Theory and Applications of Robust Optimization." In: *SIAM Review* 53.3 (2011), pages 464–501 (cited on page 17).

[BBL14]   C. Bliek, P. Bonami, and A. Lodi. "Solving Mixed-Integer Quadratic Programming problems with IBM-CPLEX: A Progress Report." In: *Proceedings of the Twenty-Sixth RAMP Symposium*. 2014, pages 171–180 (cited on pages 17, 21, 28).

[BD02]   I. M. Bomze and E. De Klerk. "Solving Standard Quadratic Optimization Problems via Linear, Semidefinite and Copositive Programming." In: *Journal of Global Optimization* 24.2 (2002), pages 163–185 (cited on page 35).

[BEG94]   A. Ben-Tal, G. Eiger, and V. Gershovitz. "Global minimization by reducing the duality gap." In: *Mathematical Programming* 63.1 (1994), pages 193–212 (cited on pages 87, 95, 131).

[BEJ16]   C. Brás, G. Eichfelder, and J. Júdice. "Copositivity tests based on the linear complementarity problem." In: *Computational Optimization and Applications* 63 (2016), pages 461–493 (cited on page 34).

[Bel+09]   P. Belotti, J. Lee, L. Liberti, F. Margot, and A. Wächter. "Branching and Bounds Tightening techniques for Non-Convex MINLP." In: *Optimization Methods and Software* 24 (2009), pages 597–634 (cited on page 17).

[Bel+13]   P. Belotti, C. Kirches, S. Leyffer, J. Linderoth, J. Luedtke, and A. Mahajan. "Mixed-Integer Nonlinear Optimization." In: *Acta Numerica* 22 (2013), pages 1–131 (cited on page 16).

[Ben+04]   A. Ben-Tal, A. Goryashko, E. Guslitzer, and A. Nemirovski. "Adjustable robust solutions of uncertain linear programs." In: *Mathematical Programming* 99.2 (2004), pages 351–376 (cited on pages 20, 149).

[Ber+12]   T. Berthold, A. M. Gleixner, S. Heinz, and S. Vigerske. "Analyzing the computational impact of MIQCP solver components." In: *Numerical Algebra, Control and Optimization* 2.4 (2012), pages 739–748 (cited on page 21).

[Ber14]   T. Berthold. "Heuristic algorithms in global MINLP solvers." PhD thesis. Technische Universität Berlin, 2014 (cited on pages 17, 47).

[BG14]   T. Berthold and A. M. Gleixner. "Undercover: a primal MINLP heuristic exploring a largest sub-MIP." In: *Mathematical Programming* 144.1 (2014), pages 315–346 (cited on page 17).

[BGL16]   P. Bonami, O. Günlük, and J. Linderoth. *Solving box-constrained nonconvex quadratic programs.* Optimization Online. 2016. URL: http://www.optimization-online.org/DB_HTML/2016/06/5488.html (visited on 07/29/2016) (cited on pages 28, 81).

[BGN09]   A. Ben-Tal, L. E. Ghaoui, and A. Nemirovski. *Robust Optimization.* Princeton Series in Applied Mathematics. Princeton University Press, 2009 (cited on pages 4, 17, 148).

[Bix+]   B. Bixby, M. Fenelon, Z. Gu, E. Rothberg, and R. Wunderling. *One Size Fits All?: Computational Tradeoffs in Mixed Integer Programming Software.* URL: http://slideplayer.com/slide/4587006/ (visited on 10/27/2016) (cited on page 170).

[Bix+00]   E. R. Bixby, M. Fenelon, Z. Gu, E. Rothberg, and R. Wunderling. "MIP: Theory and Practice — Closing the Gap." In: *System Modelling and Optimization: Methods, Theory and Applications. 19th IFIP TC7 Conference on System Modelling and Optimization July 12–16, 1999, Cambridge, UK.* Edited by M. J. D. Powell and S. Scholtes. Boston, MA: Springer US, 2000, pages 19–49 (cited on page 1).

[BKR15]   N. Boland, T. Kalinowski, and F. Rigterink. "A polynomially solvable case of the pooling problem." In: *arXiv* 1508.03181v2 (2015) (cited on page 87).

[BKR16]   N. Boland, T. Kalinowski, and F. Rigterink. "New multi-commodity flow formulations for the pooling problem." In: *Journal of Global Optimization* 66.4 (2016), pages 669–710 (cited on page 87).

[BL85]   T. E. Baker and L. S. Lasdon. "Successive Linear Programming at Exxon." In: *Management Science* 31 (1985), pages 264–274 (cited on page 87).

[BLT08]   I. M. Bomze, M. Locatelli, and F. Tardella. "New and old bounds for standard quadratic optimization: dominance, equivalence and incomparability." In: *Mathematical Programming* 115.1 (2008), pages 31–64 (cited on page 35).

[BMN11]   P. Belotti, A. J. Miller, and M. Namazifar. "Linear Inequalities for Bounded Products of Variables." In: *SIAG/OPT Views-and-News* 22.1 (2011), pages 1–8 (cited on pages 16, 27).

[BN99]   A. Ben-Tal and A. Nemirovski. "Robust Solutions of Uncertain Linear Programs." In: *Operations Research Letters* 25.1 (1999), pages 1–13 (cited on page 18).

[Bol+15] N. Boland, T. Kalinowski, F. Rigterink, and M. Savelsbergh. *A special case of the generalized pooling problem arising in the mining industry.* Optimization Online. 2015. URL: http://www.optimization-online. org/DB_HTML/2015/07/5025.html (visited on 07/29/2016) (cited on pages 3, 85, 87).

[Bol+16] N. Boland, S. S. Dey, T. Kalinowski, M. Molinaro, and F. Rigterink. "Bounding the gap between the McCormick relaxation and the convex hull for bilinear functions." In: *Mathematical Programming* (2016), pages 1–13 (cited on page 26).

[Bom+00] I. Bomze, M. Dür, E. de Klerk, C. Roos, A. Quist, and T. Terlaky. "On Copositive Programming and Standard Quadratic Optimization Problems." In: *Journal of Global Optimization* 18.4 (2000), pages 301–320 (cited on pages 34, 35).

[Bom98]   I. M. Bomze. "On Standard Quadratic Optimization Problems." In: *Journal of Global Optimization* 13.4 (1998), pages 369–387 (cited on pages 2, 34, 35).

[Bon]   P. Bonami. *Algorithms for MINLP, Presentation at CO@Work 2015, ZIB, Berlin, Germany*. URL: http://co-at-work.zib.de/files/pierre_minlp.pdf (visited on 02/02/2016) (cited on page 28).

[Bon+16]   P. Bonami, A. Lodi, J. Schweiger, and A. Tramontani. *Solving Standard Quadratic Programming by Cutting Planes*. Technical report DS4DM-2016-001. Polytechnique Montréal, June 2016 (cited on pages 6, 34).

[BS03]   D. Bertsimas and M. Sim. "Robust Discrete Optimization and Network Flows." In: *Mathematical Programming* 98 (2003), pages 49–71 (cited on page 18).

[BS04]   D. Bertsimas and M. Sim. "The Price of Robustness." In: *Operations Research* 52.1 (2004), pages 35–53 (cited on page 19).

[BST09]   X. Bao, N. Sahinidis, and M. Tawarmalani. "Multiterm Polyhedral Relaxations for Nonconvex, Quadratically Constrained Quadratic Programs." In: *Optimization Methods & Software* 24 (2009), pages 485–504 (cited on pages 16, 27).

[BV04]   S. Boyd and L. Vandenberghe. *Convex Optimization*. New York, NY, USA: Cambridge University Press, 2004 (cited on pages 13, 21).

[BV08]   S. Burer and D. Vandenbussche. "A finite branch-and-bound algorithm for nonconvex quadratic programming via semidefinite relaxations." In: *Mathematical Programming* 113.2 (2008), pages 259–282 (cited on page 23).

[BV11]   M. Bussieck and S. Vigerske. "MINLP Solver Software." In: *Wiley Encyclopedia of Operations Research and Management Science*. John Wiley & Sons, Inc., 2011 (cited on page 17).

[Cap+16]   A. Caprara, M. Carvalho, A. Lodi, and G. J. Woeginger. "Bilevel Knapsack with Interdiction Constraints." In: *INFORMS Journal on Computing* 28.2 (2016), pages 319–333 (cited on page 44).

[CKM16]   F. Ceccon, G. Kouyialis, and R. Misener. "Using functional programming to recognize named structure in an optimization problem: Application to pooling." In: *AIChE Journal* (2016) (cited on pages 85, 171).

[CL12]   A. Costa and L. Liberti. "Relaxations of Multilinear Convex Envelopes: Dual is Better than Primal." In: *Experimental Algorithms: LNCS*. Edited by R. Klasing. Volume 7276. Springer, 2012, pages 87–98 (cited on page 16).

[CN06]   G. Csardi and T. Nepusz. "The igraph software package for com-
             plex network research." In: *InterJournal* Complex Systems (2006),
             page 1695. URL: http://igraph.org (cited on page 60).

[Com]   E. Commission. *Energy Roadmap 2050.* URL: https://ec.europa.
             eu/energy/sites/ener/files/documents/2012_energy_roadmap_
             2050_en_0.pdf (visited on 09/27/2016) (cited on page 4).

[Cop+99]   D. Coppersmith, O. Günlük, J. Lee, and J. Leung. *A Polytope for
             a Product of Real Linear Functions in 0/1 Variables.* Technical report
             RC21568. IBM Watson Research, 1999 (cited on page 26).

[Cou]   Couenne, an exact solver for nonconvex MINLPs. URL: https://
             projects.coin-or.org/Couenne (cited on pages 1, 17).

[CPT99]   A. Caprara, D. Pisinger, and P. Toth. "Exact Solution of the Quadratic
             Knapsack Problem." In: *INFORMS Journal on Computing* 11.2 (1999),
             pages 125–137 (cited on page 80).

[Cre]   Creative Commons. *Attribution 3.0 Unported License.* URL: http:
             //creativecommons.org/licenses/by/3.0/ (visited on 06/09/2017)
             (cited on pages 158, 159).

[DAm+10]   C. D'Ambrosio, A. Frangioni, L. Liberti, and A. Lodi. "Experiments
             with a Feasibility Pump Approach for Nonconvex MINLPs." In:
             *Experimental Algorithms: 9th International Symposium, SEA 2010, Ischia
             Island, Naples, Italy, May 20-22, 2010. Proceedings.* Berlin, Heidelberg:
             Springer Berlin Heidelberg, 2010, pages 350–360 (cited on page 17).

[DG15]   S. S. Dey and A. Gupte. "Analysis of MILP Techniques for the
             Pooling Problem." In: *Operations Research* 63.2 (2015), pages 412–427
             (cited on page 87).

[DLL11]   C. D'Ambrosio, J. T. Linderoth, and J. Luedtke. "Valid Inequalities for
             the Pooling Problem with Binary Variables." In: *Integer Programming
             and Combinatorial Optimization, 15th International IPCO Conference
             Proceedings.* Springer, 2011, pages 117–129 (cited on page 87).

[DM02]   E. D. Dolan and J. J. Moré. "Benchmarking optimization software
             with performance profiles." English. In: *Mathematical Programming*
             91.2 (2002), pages 201–213 (cited on page 69).

[Dür10]   M. Dür. "Recent Advances in Optimization and its Applications in Engineering: The 14th Belgian-French-German Conference on Optimization." In: Berlin, Heidelberg: Springer Berlin Heidelberg, 2010. Chapter Copositive Programming – a Survey, pages 3–20 (cited on pages 2, 34).

[FA90]   C. A. Floudas and A. Aggarwal. "A Decomposition Strategy for Global Optimum Search in the Pooling Problem." In: *ORSA Journal on Computing* 2 (1990), pages 225–235 (cited on page 87).

[FHJ92]   L. Foulds, D. Haugland, and K. Jornsten. "A Bilinear Approach to the Pooling Problem." In: *Optimization* 24 (1992), pages 165–180 (cited on pages 87, 131).

[FLM13]   M. Fampa, J. Lee, and W. Melo. *On global optimization with indefinite quadratics*. Preprint NI13066. Isaac Newton Institute, 2013 (cited on pages 28, 29).

[Füg+11]   A. Fügenschuh, B. Hiller, J. Humpola, T. Koch, T. Lehman, R. Schwarz, J. Schweiger, and J. Szabó. "Gas Network Topology Optimization for Upcoming Market Requirements." In: *IEEE Proceedings of the 8th International Conference on the European Energy Market (EEM), 2011* (2011), pages 346–351 (cited on pages 139, 144).

[GAMa]   GAMS. URL: www.gams.com (cited on page 130).

[GAMb]   GAMS Model Library. URL: http://www.gams.com/modlib/modlib.htm (cited on page 131).

[Gasa]   Gaslib. *A library of gas network instances*. URL: http://gaslib.zib.de (cited on pages 5, 158, 170).

[Gasb]   Gaslib. *Data*. URL: http://gaslib.zib.de/data.html (visited on 02/08/2017) (cited on page 158).

[Gasc]   Gaslib. *Test Data*. URL: http://gaslib.zib.de/testData.html (visited on 02/08/2017) (cited on page 159).

[Geo72]   A. Geoffrion. "Generalized Benders Decomposition." In: *Journal of Optimization Theory and Applications* 10.4 (1972), pages 237–260 (cited on page 150).

[GG98]   B. Galan and I. E. Grossmann. "Optimal Design of Distributed Wastewater Treatment Networks." In: *Industrial & Engineering Chemistry Research* 37.10 (1998), pages 4036–4048 (cited on page 85).

[GHS80]   G. Gallo, P. Hammer, and B. Simeone. "Quadratic knapsack problems." In: *Combinatorial Optimization*. Edited by M. Padberg. Volume 12. Mathematical Programming Studies. Springer Berlin Heidelberg, 1980, pages 132–149 (cited on page 80).

[Gib+97]   L. Gibbons, D. Hearn, P. Pardalos, and M. Ramana. "Continuous Characterizations of the Maximum Clique Problem." In: *Mathematics of Operations Research* 22.3 (1997), pages 754–768 (cited on page 83).

[Gle+16]   A. M. Gleixner, T. Berthold, B. Müller, and S. Weltge. *Three Enhancements for Optimization-Based Bound Tightening*. Optimization Online. 2016. URL: http://www.optimization-online.org/DB_HTML/2016/03/5356.html (visited on 08/09/2016) (cited on page 17).

[GMF09]   C. E. Gounaris, R. Misener, and C. A. Floudas. "Computational Comparison of Piecewise-Linear Relaxations for Pooling Problems." In: *Industrial & Engineering Chemistry Research* 48.12 (2009), pages 5742–5766 (cited on page 87).

[GMM14]   B. Geißler, A. Martin, and A. Morsi. *Lamatto++*. Information available at http://www.mso.math.fau.de/edom/projects/lamatto.html. 2014 (cited on page 157).

[GR85]   O. K. Gupta and A. Ravindran. "Branch and Bound Experiments in Convex Nonlinear Integer Programming." In: *Management Science* 31 (1985), pages 1533–1546 (cited on page 14).

[Gup+15]   A. Gupte, S. Ahmed, S. Dey, and M. Cheon. "Pooling problems: an overview." In: *Optimization and Analytics in the Oil and Gas Industry* accepter (2015) (cited on page 87).

[Gup12]   A. Gupte. "Mixed Integer Bilinear Programming with Applications to the Pooling Problem." PhD thesis. Georgia Institute of Technology, 2012 (cited on pages 87, 88).

[GYH15]   B. L. Gorissen, İ. Yanıkoğlu, and D. den Hertog. "A practical guide to robust optimization." In: *Omega* 53 (2015), pages 124–137 (cited on page 17).

[Hau15]   D. Haugland. "The computational complexity of the pooling problem." In: *Journal of Global Optimization* 64.2 (2015), pages 199–215 (cited on page 87).

[Hav78]   C. A. Haverly. "Studies of the Behavior of the Recursion for the Pooling Problem." In: *SIGMAP Bulletin* 25 (1978), pages 19–28 (cited on pages 85, 87, 92, 131).

[Hav79]   C. A. Haverly. "Behaviour of the Recursion Model - More Studies."
            In: *SIGMAP Bulletin* 26 (1979), pages 12–28 (cited on page 131).

[HPT00]   H. Horst, P. M. Pardalos, and V. Thoai. *Introduction to Global Opti-
            mization*. 2nd. Dordrecht: Kluwer, 2000 (cited on page 13).

[Hum+15]   J. Humpola, I. Joormann, D. Oucherif, M. E. Pfetsch, L. Schewe, M.
            Schmidt, and R. Schwarz. *GasLib – A Library of Gas Network Instances*.
            Technical report. Nov. 2015. URL: http://www.optimization-
            online.org/DB_HTML/2015/11/5216.html (visited on 06/09/2017)
            (cited on page 158).

[Hum14]   J. Humpola. "Gas Network Optimization by MINLP." PhD thesis.
            Technische Universität Berlin, 2014 (cited on pages 139, 144, 167).

[IBMa]   IBM CPLEX 12.6.3 User's Manual. *Boolean Quadric Polytope (BQP)
            cuts*. URL: http://www-01.ibm.com/support/knowledgecenter/
            SSSA5P_12.6.3/ilog.odms.cplex.help/CPLEX/UsrMan/topics/
            discr_optim/mip/cuts/28_BQPcuts.html (visited on 02/02/2016)
            (cited on pages 28, 81).

[IBMb]   IBM CPLEX Optimizer. URL: http://www.cplex.com (cited on
            pages 2, 17, 31, 48).

[Kar72]   R. M. Karp. "Reducibility among Combinatorial Problems." In:
            Complexity of Computer Computations. Edited by R. E. Miller, J. W.
            Thatcher, and J. D. Bohlinger. The IBM Research Symposia Series.
            Springer US, 1972, pages 85–103 (cited on pages 35, 46).

[Koc+15]   T. Koch, B. Hiller, M. E. Pfetsch, and L. Schewe, editors. *Evaluating
            Gas Network Capacities*. SIAM-MOS series on Optimization. SIAM,
            2015. xvi + 364 (cited on pages 6, 139, 143, 145, 157).

[KTK80]   M. Kozlov, S. Tarasov, and L. Khachiyan. "The polynomial solvability
            of convex quadratic programming." In: *USSR Computational Mathe-
            matics and Mathematical Physics* 20.5 (1980), pages 223–228 (cited on
            page 21).

[LD60]   A. H. Land and A. G. Doig. "An Automatic Method for Solving Dis-
            crete Programming Problems." In: *Econometrica* 28 (1960), pages 497–
            520 (cited on page 13).

[Lie+09]   C. Liebchen, M. Lübbecke, R. Möhring, and S. Stiller. "The Concept of Recoverable Robustness, Linear Programming Recovery, and Railway Applications." In: *Robust and Online Large-Scale Optimization: Models and Techniques for Transportation Systems*. Edited by R. K. Ahuja, R. H. Möhring, and C. D. Zaroliagis. Springer Berlin Heidelberg, 2009, pages 1–27 (cited on pages 20, 149).

[LM06]   L. Liberti and N. Maculan, editors. *Global Optimization – From Theory to Implementation*. Volume 84. Nonconvex Optimization and Its Applications. Springer US, 2006 (cited on page 13).

[LNL12]   J. Luedtke, M. Namazifar, and J. T. Linderoth. "Some Results on the Strength of Relaxations of Multilinear Functions." In: *Mathematical Programming, Series B* 136 (2012), pages 325–351 (cited on page 26).

[Lot+16]   I. Lotero, F. Trespalacios, I. E. Grossmann, D. J. Papageorgiou, and M.-S. Cheon. "An MILP-MINLP decomposition method for the global optimization of a source based model of the multiperiod blending problem." In: *Computers & Chemical Engineering* 87 (2016), pages 13–35 (cited on page 87).

[LRW14]   A. Lodi, T. K. Ralphs, and G. J. Woeginger. "Bilevel programming and the separation problem." In: *Mathematical Programming* 146.1-2 (2014), pages 437–458 (cited on page 44).

[LS13]   M. Locatelli and F. Schoen. *Global optimization: theory, algorithms, and applications*. Volume 15. MPS-SIAM Series on Optimization. SIAM, 2013 (cited on page 13).

[LTB11]   X. Li, A. Tomasgard, and P. Barton. "Nonconvex Generalized Benders Decomposition for Stochastic Separable Mixed-Integer Nonlinear Programs." In: *Journal of Optimization Theory and Applications* 151.3 (2011), pages 425–454 (cited on page 150).

[LTB16]   X. Li, A. Tomasgard, and P. I. Barton. "Natural gas production network infrastructure development under uncertainty." In: *Optimization and Engineering* (2016), pages 1–28 (cited on page 150).

[McC76]   G. P. McCormick. "Computability of Global Solutions to Factorable Nonconvex Programs: Part I—Convex Underestimating Problems." In: *Mathematical Programming* 10 (1976), pages 147–175 (cited on pages 2, 15, 16, 23–25).

[MF05]   C. A. Meyer and C. A. Floudas. "Convex Envelopes for Edge-Concave Functions." In: *Mathematical Programming, Series B* 103 (2005), pages 207–224 (cited on pages 16, 27).

[MF06]   C. A. Meyer and C. A. Floudas. "Global optimization of a combinatorially complex generalized pooling problem." In: *AIChE Journal* 52.3 (2006), pages 1027–1037 (cited on page 87).

[MF10]   R. Misener and C. A. Floudas. "Global Optimization of Large-Scale Generalized Pooling Problems: Quadratically Constrained MINLP Models." In: *Industrial & Engineering Chemistry Research* 49.11 (2010), pages 5424–5438 (cited on page 87).

[MF13]   R. Misener and C. A. Floudas. "GloMIQO: Global mixed-integer quadratic optimizer." In: *Journal of Global Optimization* 57.1 (2013), pages 3–50 (cited on pages 17, 29).

[MF14]   R. Misener and C. A. Floudas. "ANTIGONE: Algorithms for coNTinuous / Integer Global Optimization of Nonlinear Equations." In: *Journal of Global Optimization* 59.2 (2014), pages 503–526 (cited on page 17).

[MGF10]   R. Misener, C. E. Gounaris, and C. A. Floudas. "Mathematical modeling and global optimization of large-scale extended pooling problems with the (EPA) complex emissions constraints." In: *Computers & Chemical Engineering* 34.9 (2010), pages 1432–1456 (cited on page 87).

[MS65]   T. S. Motzkin and E. G. Straus. "Maxima for graphs and a new proof of a theorem of Turán." In: *Canadian Journal of Mathematics* 17 (1965), pages 533–540 (cited on pages 2, 34, 35).

[MSF15]   R. Misener, J. B. Smadbeck, and C. A. Floudas. "Dynamically-Generated Cutting Planes for Mixed-Integer Quadratically-Constrained Quadratic Programs and their Incorporation into GloMIQO 2.0." In: *Optimization Methods and Software* 30 (2015), pages 215–249 (cited on pages 17, 26, 27, 30).

[Nes99]   Y. Nesterov. *Global quadratic optimization on the sets with simplex structure.* CORE Discussion Papers; 1999/15. 1999 (cited on page 35).

[Pfe+14]   M. E. Pfetsch, A. Fügenschuh, B. Geißler, N. Geißler, R. Gollmer, B. Hiller, J. Humpola, T. Koch, T. Lehmann, A. Martin, A. Morsi, J. Rövekamp, L. Schewe, M. Schmidt, R. Schultz, R. Schwarz, J. Schweiger, C. Stangl, M. C. Steinbach, S. Vigerske, and B. M. Willert. "Validation of nominations in gas network optimization: models, methods, and solutions." In: *Optimization Methods and Software* (2014) (cited on pages 139, 143).

[PV92]   P. M. Pardalos and S. A. Vavasis. "Quadratic Programming with One Negative Eigenvalue is NP-hard." In: *Journal of Global Optimization* 1 (1992), pages 15–22 (cited on page 21).

[QG95]   I. Quesada and I. Grossmann. "Global optimization of bilinear process networks with multicomponent flows." In: *Computers & Chemical Engineering* 19.12 (1995), pages 1219–1242 (cited on pages 3, 85, 87).

[RC09]   R. Misener and C.A. Floudas. "Advances for the Pooling Problem: Modeling, Global Optimization, & Computational Studies." In: *Applied and Computational Mathematics* 8 (2009), pages 3–22 (cited on page 87).

[Rik97]   A. D. Rikun. "A Convex Envelope Formula for Multilinear Functions." In: *Journal of Global Optimization* 10 (1997), pages 425–437 (cited on page 26).

[Roc70]   R. T. Rockafellar. *Convex Analysis.* Princeton University Press, 1970 (cited on pages 13, 15).

[Rui+12] M. Ruiz, O. Briant, J.-M. Clochard, and B. Penz. "Large-scale standard pooling problems with constrained pools and fixed demands." In: *Journal of Global Optimization* 56.3 (2012), pages 939–956 (cited on page 87).

[Rus06]   A. P. Ruszczyński. *Nonlinear optimization.* Volume 13. Princeton university press, 2006 (cited on page 12).

[SA92]   H. D. Sherali and A. R. Alameddine. "A new reformulation-linearization technique for bilinear programming problems." In: *Journal of Global Optimization* 2.4 (1992), pages 379–410 (cited on pages 2, 29).

[SA94]   H. D. Sherali and W. P. Adams. "A hierarchy of relaxations and convex hull characterizations for mixed-integer zero-one programming problems." In: *Discrete Applied Math.* 52 (1994), pages 83–106 (cited on page 30).

[Sah74]   S. Sahni. "Computationally Related Problems." In: *SIAM Journal on Computing* 3.4 (1974), pages 262–279 (cited on page 21).

[Sah96]   N. V. Sahinidis. "BARON: A General Purpose Global Optimization Software Package." In: *Journal of Global Optimization* 8 (1996), pages 201–205 (cited on page 17).

[SBL10]   A. Saxena, P. Bonami, and J. Lee. "Convex relaxations of non-convex mixed integer quadratically constrained programs: projected formulations." In: *Mathematical Programming* 130.2 (2010), pages 359–413 (cited on page 28).

[Sch16]   J. Schweiger. "Gas Network Extension Planning for Multiple Demand Scenarios." In: *Operations Research Proceedings 2014: Selected Papers of the Annual International Conference of the German Operations Research Society (GOR), RWTH Aachen University, Germany, September 2-5, 2014.* Edited by M. Lübbecke, A. Koster, P. Letmathe, R. Madlener, B. Peis, and G. Walther. Cham: Springer International Publishing, 2016, pages 539–544 (cited on pages 7, 148).

[Sch17]   J. Schweiger. "Exploiting structure in non-convex quadratic optimization and gas network planning under uncertainty." PhD thesis. Technische Universität Berlin, 2017. URL: http://dx.doi.org/10.14279/depositonce-6015 (cited on pages 49, 70, 83, 133, 167).

[SCI]   SCIP: Solving Constraint Integer Programs. URL: http://scip.zib.de (cited on pages 1, 17).

[SG91]   N. V. Sahinidis and I. E. Grossmann. "Convergence properties of generalized Benders Decomposition." In: *Computers and Chemical Engineering* 15 (1991), pages 481–491 (cited on page 150).

[Spr16]   K. Spreckelsen. *Open Grid Europe GmbH, former project lead ForNe.* Personal communication. Oct. 2016 (cited on page 6).

[SPW09]   K. J. Singh, A. B. Philpott, and R. K. Wood. "Dantzig-Wolfe Decomposition for Solving Multistage Stochastic Capacity-Planning Problems." In: *Operations Research* 57.5 (2009), pages 1271–1286 (cited on pages 154, 167).

[ST08]   A. Scozzari and F. Tardella. "A clique algorithm for standard quadratic programming." In: *Discrete Applied Mathematics* 156.13 (2008), pages 2439–2448 (cited on pages 35, 49, 70).

[Sym16]   SymPy Development Team. *SymPy: Python library for symbolic mathematics.* 2016. URL: http://www.sympy.org (visited on 10/20/2016) (cited on page 111).

[Tar04]   F. Tardella. "On the existence of polyhedral convex envelopes." In: *Frontiers in Global Optimization.* Edited by C. A. Floudas and P. Pardalos. Boston, MA: Springer US, 2004, pages 563–573 (cited on pages 15, 16).

[Tar08]   F. Tardella. "Existance and sum-decomposability of vertex polyhe-
           dral convex envelopes." In: *Optimization Letters* 2 (2008), pages 363–
           375 (cited on page 16).

[TS02]   M. Tawarmalani and N. V. Sahinidis. *Convexification and Global
           Optimization in Continuous and Mixed-Integer Nonlinear Programming:
           Theory, Algorithms, Software, and Applications.* Boston MA: Kluwer
           Academic Publishers, 2002 (cited on pages 16, 87, 98).

[TS04]   M. Tawarmalani and N. V. Sahinidis. "Global Optimization of
           Mixed integer Nonlinear Programs: A Theoretical and Computa-
           tional Study." In: *Mathematical Programming* 99 (2004), pages 563–591
           (cited on page 16).

[TS05]   M. Tawarmalani and N. Sahinidis. "A Polyhedral Branch-and-Cut
           Approach to Global Optimization." In: *Mathematical Programming*
           103 (2 2005), pages 225–249 (cited on page 17).

[Vav90]   S. A. Vavasis. "Quadratic programming is in NP." In: *Information
           Processing Letters* 36.2 (1990), pages 73–77 (cited on page 21).

[VG16]   S. Vigerske and A. Gleixner. *SCIP: Global Optimization of Mixed-Integer
           Nonlinear Programs in a Branch-and-Cut Framework.* eng. Technical
           report 16-24. Zuse Institute Berlin, 2016 (cited on pages 16, 17, 157).

[Vig13]   S. Vigerske. "Decomposition of Multistage Stochastic Programs
           and a Constraint Integer Programming Approach to Mixed-Integer
           Nonlinear Programming." PhD thesis. Humboldt-Universität zu
           Berlin, 2013 (cited on pages 10, 15–17).

[Vis01]   V. Visweswaran. "MINLP: applications in blending and pooling
           problems." In: *Encyclopedia of Optimization.* Edited by C. A. Floudas
           and P. M. Pardalos. Boston, MA: Springer US, 2001, pages 1399–1405
           (cited on page 87).

[Wal10]   S. W. Wallace. "Stochastic programming and the option of doing it
           differently." In: *Annals of Operations Research* 177.1 (2010), pages 3–8
           (cited on page 154).